Contents

Foreword
vi

CHAPTER 1
Getting Started, by Enoch H. Tompkins
1

CHAPTER 2
How to Work With Bees, by Enoch H. Tompkins
12

CHAPTER 3
Installing Package Bees, by Enoch H. Tompkins
23

CHAPTER 4
Queen Supersedure, by Enoch H. Tompkins
34

CHAPTER 5
Spring Management of Overwintered Colonies,
by Enoch H. Tompkins
36

CHAPTER 6
Summer Management, by Enoch H. Tompkins
54

CHAPTER 7
Fall and Winter Management, by Enoch H. Tompkins   71

CHAPTER 8
Where Bees Can Be Kept, by the Garden Way Staff   80

CHAPTER 9
Development of the American Beehive,   85
by Spencer M. Riedel, Jr.

CHAPTER 10
**Seasonal Colony Activity and
Individual Bee Development,** by Norbert M. Kauffeld   90

CHAPTER 11
Bee Behavior, by Stephen Taber III   97

CHAPTER 12
Honey Bee Nutrition, by L. N. Standifer   111

CHAPTER 13
Managing Colonies for High Honey Yields,   119
by F. E. Moeller

CHAPTER 14
Honey in Cooking, by Roger Griffith   131

CHAPTER 15
Queen and Package Bee Production,   137
by William C. Roberts and Warren Whitcomb, Jr.

CHAPTER 16
**Identification and Control
of Honeybee Diseases,** by H. Shimanuki   144

CHAPTER 17
Controlling the Greater Wax Moth,   162
by Agricultural Research Service

CHAPTER 18

Pesticides, by Philip F. Torchio   175

CHAPTER 19

Commercial Beekeeping Equipment,   188
by Charles D. Owens and Benjamin F. Detroy

Glossary   202

Additional Sources of Information   211

Other Garden Way Books   213

Index   214

Foreword

The Garden Way family has gradually been turning to bees. More and more members of that family, already deep into gardening, have wanted something more. Beekeeping has supplied the need. The initial cost is small. Practically no space is required. The results are so sweet. And families have found beekeeping a delightful project, one in which grade-school age children could participate—and use for those school papers.

These families, and others of our readers, have increasingly expressed a desire for a Garden Way book on the subject. Their requirements were extremely specific. Some had had difficulty, after purchasing bees, in installing them in the hive. Others found it difficult to open the hive. Give us specific information—and pictures as well, they demanded.

And there were other requests, and dozens of questions. How and when should you requeen? What kind of honey should you try to produce, the bulk honey or that honey with the wax in it? What equipment do I need to get started—and later?

Enoch Tompkins, a veteran beekeeper who has taught enough classes to learn the common problems of beginners, agreed to write all of the basic information for the beginners, and the chapters he wrote provide what a person starting out with bees should understand.

We also wanted to provide information for the more advanced beekeeper, the one who is a professional, or nearly so. The most logical way to get this information was to read through the many state and federal government bulletins on beekeeping, and to take from them the most pertinent material, offering it in easy-to-find fashion.

So, here is the book we have been asked to provide—the book for both beginner and the more advanced. And with it goes a short message: Work with your bees, and they will work hard for you.

ROGER M. GRIFFITH

CHAPTER 1

Getting Started

by Enoch H. Tompkins

The best time to get started in beekeeping is during the winter months. At that time you can read up on the subject, and order your bees and equipment. In January or February you should place your order for bees and ask that they be delivered at the time when fruit trees start to bloom in your area. This date will vary according to your location. Ask an experienced beekeeper in your area when this date will likely be.

ENOCH H. TOMPKINS, author of the first seven chapters of this book, is trainer-evaluator-statistician for the Vermont Extension Service's Rural and Farm Family Rehabilitation Program of the University of Vermont College of Agriculture at Burlington, Vermont. From 1954 to 1972 he served as agricultural economist and rural sociologist with the College of Agriculture at the University of Vermont.

Mr. Tompkins is a well-known expert on beekeeping in Vermont. He has kept bees for thirty-five years and in three states: Maine, Connecticut and Vermont. While in Connecticut, he spent one summer as an apiary inspector for the State Department of Agriculture. He has given numerous talks on bees, honey and beekeeping and has written many articles on these subjects. He teaches an Evening Division course in beekeeping at the University of Vermont. He has served as president of the Eastern Apricultural Society and as president and secretary of the Vermont Beekeepers Association. He was instrumental in founding the Vermont Beekeepers Association exhibit of old beekeeping equipment at the Shelburne Museum of Shelburne, Vermont, and maintains an observation beehive there. He also collects, cleans, repairs and does research on old beekeeping equipment for the museum collection.

If you are buying package bees, you should get your order in early because the orders are filled on the basis of first-come, first-served. It is important to get your bees early in the season, in northern states, so they will have time to build their combs and raise new bees so the hive population will be large at the time of the main honeyflow. The earlier you get the bees, the more time they have for building up and making honey. Also, the better the chance they have of producing enough honey to carry them through the next winter and perhaps make some surplus honey for you the first year. Most beginners don't think about getting bees until the spring flowers start to bloom. By that time it may be too late to get the bees and to give them a chance to do their best. If you get them too late, they may not have time to store enough honey for their winter needs and may die of starvation.

How to Buy Them

There are several different ways to obtain bees. They may be bought from southern or California beekeepers in screened packages containing two to five pounds of bees and a queen. You may be able to purchase an established hive of bees from a nearby beekeeper. Or, you might have a nearby beekeeper place a swarm of bees in a hive for you. Bees may be removed from buildings or trees. However, I would recommend that this method be left to experienced beekeepers rather than beginners. It is not as easy as it may sound.

When buying an established hive of bees, make certain it has been inspected for disease by a state apiary inspector. Ask for a certificate of inspection signed by the inspector stating that it is free of disease. Don't take anyone else's word for it. Don't take a chance on losing your bees from disease right at the start. You will have enough problems without that one.

Basic Equipment

At the same time that you order your bees, you should order the equipment that will be necessary for housing and caring for them. Several beekeeping equipment suppliers advertise basic beginner's outfits. These basic kits sold for about $52.00, plus transportation charges, in 1976. They generally contain the following:

1. One standard ten-frame hive with frames and foundation for the combs in which the bees will raise their brood and store honey and pollen for food.

2. A bee veil for protecting your face from the bees.

3. A bee smoker for controlling the bees.

4. Gloves to protect your hands.

5. A hive tool for prying the hive and frames apart.

6. A feeder for feeding the bees sugar syrup until they can support themselves with available nectar.

7. A beginner's book on beekeeping.

Basic equipment needed by the beginner includes standard ten-frame hive with frames, at left; bee veil, gloves, and bee smoker, and in front, a basic book, foundation for the frames, hive tool and entrance feeder.

The bees are generally sold separately in two- to five-pound screened packages. The three-pound package is recommended over the two-pound package, since the greater number of bees is needed to keep up the temperature of the bee cluster for raising young bees. This temperature has to be maintained at between 93 and 94 degrees Fahrenheit in the brood area. Three-pound packages sold for around $25, including shipping charges, in 1976.

This basic equipment contains only that which is needed to get the bees started and to keep them going for about the first month. Additional equipment will be needed later for providing more room for the bees as they increase in number and store honey and pollen. It is wise to order this additional equipment at the start so you will have it when needed. If additional room is not provided when needed, the hive may become overcrowded and the bees may swarm. When they do, you may lose your nectar-gathering force and hence your honey crop. The remaining bees may not be able to store enough honey to supply them for the winter. You may then lose your bees from starvation and thus a good part of your investment.

Additional Equipment Needed

In addition to the basic beekeeping outfit that has been described, there are a few more items that are desirable or which will be needed during the first year.

Deep super. The most basic piece of equipment that will be needed, perhaps as soon as a month after the package bees have been installed, is a deep super for the food chamber. This is needed to provide for most of the honey stores that the bees will be needing to see them through the winter. This deep super will need to be equipped with ten brood frames and ten sheets of brood foundation. It will be placed on top of the hive body that the bees were installed in originally. It should be added as soon as the bees have drawn out the foundation into comb in the first hive body and filled it with brood, pollen and syrup.

Queen excluder. The queen excluder is a sort of fence that has spaces large enough to permit worker bees to pass through but not the queen, which is larger than the workers. It is placed above the food chamber super to keep the queen from laying eggs in shallow supers when you are producing cut-comb or chunk-comb honey. Comb in which brood has been reared will be dark colored and its appearance will be impaired. The welded-wire excluder is recommended over the perforated-zinc excluder. The latter type has sharp edges which wear the workers' wings as they pass through.

Shallow supers. If your first year turns out to be a good one, you will need one or more shallow supers. You should have at least one; perferably three. If you have just one, and the bees need still more room for storing nectar and honey, you can remove just one frame at

A three-pound package of bees. Screen sides keep bees in package.

Queen excluder permits worker bees to go through, but prevents queen from moving from one super to another.

a time as it is filled and capped. Then you can cut out the honeycomb from the frame, add a new sheet of foundation wax, and replace the frame in the super. This should be done the same day so the bees won't start building comb in the space left by the missing frame.

If you have three shallow supers, you can keep adding them as they are needed and not have to rush so much to remove and replace one frame at a time. These will provide the bees with extra storage space so you can wait until an entire super is filled and capped before removing any honey. Nectar may contain about 80 percent water; the finished product, honey, contains about 17 percent. To evaporate the surplus water from the nectar, the bees need plenty of storage space to hold the product during the processing.

Each shallow super will need ten shallow frames and ten sheets of foundation.

Granulated sugar. For each package of bees, you need a minimum of ten pounds of granulated sugar for making syrup for feeding the bees. They will need to be fed syrup while they are building their combs and until nectar becomes plentiful.

Proper clothing. When working with bees, it is best to wear smooth, light-colored clothes. Tan, khaki-colored or white shirt and trousers or a coverall suit are good. The advantage of a coverall suit is that it may be slipped on over other clothing and provide an attire that is inconspicuous to the bees.

Bees can differentiate certain colors, and dark colors (especially blue) can be seen very clearly. If the bees get angry, they can see the dark clothing and know where to attack. Light-colored clothing will give the beginner more protection from the bees' stings and will also be cooler in warm weather.

Bee brush. A brush for brushing bees off combs, or other places where they are not wanted, is a handy tool. It may also be used to brush a swarm off the side of a tree, a fence post or the side of a building. They may be brushed into a box and dumped in front of a hive.

The bee brush is an effective tool for moving bees without hurting them.

Bee escape. A bee escape will be useful for removing bees from supers of honey that are ready for removal from the hive. It is a device for inserting into the hole in the inner cover which is then placed under the super from which the bees are to be removed. The bees go down through the hole in the bee escape and pass between two flat wire springs. If the bees try to return, the springs close and prevent them from returning to the super.

Bee escape fits into hole in inner cover, and permits bees to go down through but not return.

Second-Year Equipment

For the second year, additional equipment will be needed. Most of the first year will be spent by the bees in building combs, increasing in numbers, and storing food for winter. You probably won't get any honey for yourself the first year. In the northern states, what honey the bees make will probably be needed by the bees for winter stores. The second year should be your big honey-crop year. The combs will all be built in the brood and food chambers. You hope there will be a good supply of honey and pollen so the queen can get an early start with raising young bees. These will build up the number of bees for gathering nectar and making a good honey crop. For this reason, you will need the following additional equipment:

Shallow supers. For storing your honey crop, you will need two shallow supers for each first-year hive in addition to the one you bought for the first year, or three in all. These will need ten frames each and ten sheets of the appropriate foundation. Most hive parts are cheaper if bought in lots of five (or fifty in the case of frames). For this reason, you may wish to buy five shallow supers the first year, in anticipation of future needs. Foundation is also cheaper per sheet if bought in larger lots.

Extra hive. A complete extra hive is needed for each first-year hive, in case your bees decide to swarm. Your honey crop will be larger if you can keep them from swarming, but that will be a hard thing for a beginner to accomplish. So, it will be well to be prepared for swarming so you won't have to run to the nearest dealer in beekeeping equipment for the hive when it is needed. Your swarm probably will decide to leave while you are getting the hive and putting it together. Then you will have lost your swarm, which includes most of your field bees (the ones which would be gathering the nectar for your honey crop). Losing them will mean that you have lost most of your surplus honey for that year.

Deep super. If you decide to keep the swarm as a separate colony, you will need an extra super for the food chamber. This should be a deep super with ten brood frames and ten sheets of brood foundation.

Equipment for Extracted
Honey Production

If you decide to produce extracted honey (honey removed from the comb), you will need still more equipment. In producing extracted honey, the honey is removed from the combs by centrifugal force and the combs can be used over and over again. This saves the bees some work and results in larger crops of honey. However, more money must be invested in special equipment for this purpose.

Uncapping knife. For cutting off the thin layer of wax (known as the cell cappings), you will need an uncapping knife. This is a heavy, double-edged knife for slicing off a thin layer of wax from each side of the honeycomb so the honey may be removed from the cells. The knife is generally heated by immersing it in a pan of hot water. Some knives are heated electrically or by steam. If you use the simpler uncapping knife, which must be heated in hot water, it will be best to have two of them. That way, one can be heating while the other is being used for uncapping.

Capping tub. You will need some sort of tub or large pan for catching and holding the cappings as they are cut from the combs. A large dish pan or a laundry tub may be used for this purpose. Usually a wire screen is used to catch the cappings to allow the honey to drain from them. This screen may be supported by a wooden rack on the bottom of the pan or tub to allow space for the honey.

Extractor. A honey extractor will be needed to remove the honey from the uncapped combs. This consists of a square wire basket which revolves inside a metal tank. The uncapped combs are placed against each side of the wire basket. When the basket is turned, by means of a crank and gears, the honey is thrown out of the cells by centrifugal force. The honey hits the sides of the tank, runs down to the bottom and is drained out through a spigot in the bottom of the tank.

Extractors come in various sizes from two-frame to radial extractors holding 72 frames. A two- or four-frame hand-driven extractor will suffice for the beginner.

Costs may be cut by making your own extractor from plans that are available from Garden Way Publishing, Charlotte, Vermont 05445. Sometimes, used extractors may be purchased for about half the cost of a new one. Another way to cut costs is for several beekeepers to purchase an extractor cooperatively and to take turns using it. Or, perhaps the beginner can make arrangements for another beekeeper to do his extracting for a fee or in exchange for help with the extracting. Sometimes honey extractors can be rented by the day from an equipment dealer. The extractor should be washed thoroughly before use to make a more sanitary product and to prevent spread of disease to your bees from contaminated honey.

Storage tank. A honey storage tank will be needed for holding the honey from the extractor until it is put into jars or cans. Five-gallon tinned, or plastic, cans may be used for this purpose if you have only one or two colonies.

Sources of Beekeeping Equipment

There are a number of manufacturers and/or suppliers of beekeeping equipment in the United States and Canada. They also generally deal in bees and queens. You should write to several of them and ask for a free copy of their catalog. This will give you the names of their local dealers, if they have them. Also, they will let you know what is available and the prices charged. It pays to shop around for the lower prices. Following are the names and addresses of most of these suppliers:

1. Maxant Industries, P.O. Box 454, Ayer, MA 01432.

2. Sears, Roebuck and Co., Boston, MA 02215 and Philadelphia, PA 19132. (Ask for a copy of the Farm and Ranch Catalog.)

3. Montgomery Ward, Dept. FSC 76, Albany, NY 12201. (Write for the Suburban, Farm and Garden Catalog.)

4. Bee-Jay Farm, Rt. 1, Suwannee, GA 30174.

5. Forbes & Johnston, P.O. Box 212, Homerville, GA 31634.

6. Holloway & Sons, Inc., P.O. Box 1041, Sanford, FL 32771.

7. The Walter T. Kelley Co., Inc., Clarkson, KY 42726.

8. The A.I. Root Co., 623 West Liberty Street, P.O. Box E, Medina, OH 44256.

9. E & T Growers, Rt. 1, Spencerville, IN 46788.

10. Hubbard Apiaries, Onsted, MI 49265.

11. August Lotz Co., Boyd, WI 54726.

12. Dadant & Sons, Inc., Hamilton, IL 62341.

13. Johannessen Bee Supply, 551 West 400 No., St. George, UT 84770.

14. Miller's Frames, 2028 W. Sherman, Phoenix, AZ 85009.

15. Beeline Enterprises, Inc., 3629 North Caballero Road, Tucson, AZ 85705.

16. Glorybee Honey, 1635 River Road, Eugene, OR 97404.

17. Strauser Bee Supply, Box 991, Walla Walla, WA 99362.

18. F. W. Jones & Son Ltd., Bedford, PQ, Canada, J0J 1A0.

19. W. A. Chrysler & Son, Chatham, ON, Canada.

20. Hodgson Bee Supplies, Ltd., P.O. Box 297, New Westminster, BC, Canada V3L 4Y6.

USED EQUIPMENT

One way to cut the costs of getting started with bees is to purchase equipment that someone else has used for bees. Sometimes one may have used equipment given to him.

Second-hand hives, and hive parts should be cleaned and sterilized before being used again for bees. The reason for this is that they may harbor bee diseases. The most feared of these is American Foulbrood. This is a disease that infects and kills bee brood while in the larval or pupal stage. More will be said about it in a later section, but it is mentioned here to warn beginners of the danger of infecting their bees with this dread disease through the use of second-hand equipment. The bacteria that causes American Foulbrood can remain viable for 50 or more years on infected equipment. Don't take anyone's word that the equipment is free of disease. The organism cannot be seen without the aid of a microscope and the person who lost his bees may not have recognized the disease. The used equipment may have been lying around for years and may have passed through many hands. In the process, the history of the equipment may have become fuzzy or lost entirely. Don't take a chance on losing your bees from disease. *Sterilize the equipment!*

First, cut out the combs and scrape off all beeswax and propolis (bee glue) from all the hive parts. Dig a hole in the ground two feet square and two feet deep. Build a fire in the hole and burn all the combs and scrapings. Then fill in the hole with the soil.

How to sterilize. The equipment may be sterilized by one of several ways. Everything but the frames may be scorched with a blow torch until it is a dark brown. Or, equipment may be dipped into boiling lye water. Lye water may be prepared by adding a one-pound can of household lye to ten gallons of cool water. The water is brought to a boil before dipping the equipment. Scrubbing with a brush helps the cleaning process. Be careful when using lye because it is highly corrosive and will burn the skin if the solution is accidentally splashed on it. Use iron or galvanized iron since the lye solution will dissolve aluminum and some other metals.

Wear rubber gloves and protective goggles when working with lye to avoid burns. After the equipment has been dipped for about five minutes in the boiling lye water, it should be removed and rinsed off with plain water to remove the lye. After it dries, it is ready for use.

A newer method of disinfecting equipment is available in some states. This involves the use of ethylene oxide gas in a specially constructed chamber. Where equipment for using this is available, it is the best method, since even the combs may be sterilized.

Contact your state apiary inspector for further information and advice on dealing with suspected disease or diseased equipment.

CHAPTER 2

How to Work With Bees

by Enoch H. Tompkins

When working with bees, there are some things that you can do to cut down on the chances of unduly upsetting the bees and of being stung. Here are some simple rules to follow to accomplish this:

Work with bees only on warm sunny days, preferably when the temperature is above 70 degrees Fahrenheit and between the hours of 10 A.M. and 4 P.M. When these conditions prevail, many of the bees are away from the hive and are not there to get in your way or to challenge your actions. Bees are influenced by the weather and are likely to be bad tempered during cool, cloudy or rainy weather, or even after a rainy spell.

How to Dress

Change your clothing before working with bees if you have been working with horses, cattle or other animals. The bees don't like these odors.

It is preferable to dress in white or tan clothing when working with bees. You are less conspicuous to them in attire of that sort. Avoid blue, green or dark-colored clothing because the bees see these colors best and will be able to find you more easily if they get irritated.

Wear white socks and tuck your trousers or slacks into your socks. Bees are likely to attack your ankles because they are on the level of the hive entrance. Occasionally a bee may climb up your leg if you don't have your trousers tucked into your socks or closed by a rubber band, string or other means. When she gets squeezed by your trousers she is likely to retaliate with her sting.

Enoch Tompkins, wearing the coveralls favored by professional beekeepers, checks fastening of his bee vail to make certain bees can't enter under it.

Always wear your bee veil when working with bees. That will keep the bees away from your face if properly worn. Wear it with a fairly broad-brimmed hat, preferably of light colored straw. Avoid felt hats because the bees dislike the animal odor of felt. Make certain that the top of the veil fits snugly around the hat crown just above the brim so bees can't get in at that point. Check the lower skirt of the veil to make sure it is snug around the upper part of your chest so bees can't enter there. Turn up your shirt collar. That will help to keep the back of the veil away from your head and neck. Ask your local bee equipment dealer or an experienced beekeeper how to put your veil on properly. You may think this is a simple matter, but I have seen some weird and unprotective ways of wearing a bee veil by beginners. It won't give you much protection if it isn't worn properly.

Wear your bee gloves when you first start working with bees. They will protect your hands from stings and give you confidence at the start. Later, when you have become more accustomed to the bees, you will want to try working without them. They are quite cumbersome and wearing them makes it more difficult to handle the frames of the hive. They also make your hands and arms uncomfortably warm when the outside temperature climbs into the 80's and 90's. Keep them handy, though, so you can pull them on if the bees get too cantankerous.

Use of Smoker

Before approaching the hive, make certain that your smoker is lit and smoldering nicely. For fuel, you may use clean burlap, dry wood shavings, pine needles, or dry rotted wood. Start the fire by lighting a small wad of paper and dropping it into the bottom of the smoker canister. Then gradually add your fuel, while slowly working the bellows. Work the bellows occasionally while you are working with the bees to make certain the fire doesn't go out. This is likely to happen just at the time you need it most, when the bees get upset and need to be quieted.

Now that you are properly equipped, it is time to approach the hive. When working with bees, it is best to stand at one side of the hive. Standing by the side of the hive makes it easier to lift out the frames containing the combs. Don't stand in front of the entrance to the hive because the bees don't like to be obstructed in their flight. If they bump into you they are more likely to sting. If you stand in front of the hive, you may disorient them and delay their entrance into the hive with their loads of nectar and pollen. If possible, stand with your back to the sun. This makes it easier to see into the cells of the brood comb and to distinguish the eggs and very young larvae.

LIGHTING THE SMOKER

Matches are always available in metal box fastened to smoker (left). Start fire with wad of paper placed in smoker canister (below left). Slowly work bellows, and add fuel, such as these wood shavings (below, right).

Opening the Hive

When you are ready to open the hive, blow two or three puffs of smoke into the entrance. This disorganizes the guards and causes the bees to fill up with honey. When they are full of honey, they are in a better mood and easier to work with. Wait a minute or two, then remove the outer cover of the hive. Place the outer cover upside down on the ground behind the hive. Blow a couple of puffs of smoke into the hole of the inner cover. Next, use the flat end of your hive tool to pry up one corner of the inner cover. Blow some smoke through the crack as you lift off the inner cover. Place the inner cover upside down in front of the hive with the end resting on the entrance. This permits the bees to crawl back into the hive. If there are supers on the hive, set these crosswise on the overturned cover and check the brood chamber first. Direct a couple puffs of smoke across the tops of the frames to drive the bees down out of the way. Now you are ready to start removing combs for inspection.

Generally, it is best first to remove the comb nearest you at the side of the hive. Loosen the frame from the side of the hive by inserting the bent end of your hive tool between the top bar of the frame (near one end) and the side of the hive. Apply pressure against the other end of the tool to pry the frame away from the side of the hive. Repeat the operation at the other end of the frame. Next do the same thing between the first and the second frames to separate them. Now, insert the bent end of your hive tool between the first and second frames near one end and turn it a little so the corner of the tool digs into the first frame. Pry up on the frame to lift it enough so you can grasp it with your fingers. Do the same at the other end of the frame. When you have raised it enough to get a good grip with both hands, you may slowly and carefully lift the frame from the hive. Be careful not to crush any bees in the process. To do so will irritate other bees and make them want to sting you.

WORK SLOWLY

Work with slow motions since a fast motion will be more noticeable to the bees and will more likely result in stings.

The first comb will probably contain mostly honey and no brood. Lean it against the back of the hive or against the farther side. Place it in the shade so the sun won't melt the comb. Removing this comb will leave room for working with the remaining combs.

Pry the second frame from the third and remove it from the hive. Try to hold the frames above the hive as you are inspecting them so that the queen bee will fall into the hive if she should drop off the comb. Otherwise, she may become lost in the grass.

Hold the frame so the sun shines over your shoulder and down into the cells. Look into the cells to see if you can find any eggs or very young larvae. The eggs will look like small white commas attached to the bottom of the cells. These hatch into small, white, grublike larvae and will be curled up in a milky mass of royal jelly or bee milk. If you find eggs or young larvae, you know that the queen is present and laying. The exception would be laying workers and this will be discussed in another section.

Look for sealed brood and check that for uniformity of the queen's egg laying and for signs of disease. Bee diseases are discussed in a later section.

ENOUGH HONEY?

Check to make certain that the bees have enough honey. They should never have less than fifteen pounds. This would be the equivalent of three deep frames, or five shallow frames full of honey. If you find that they have less than this amount, you should feed them sugar syrup made from equal parts of granulated sugar and hot water. More about feeding will be presented under that heading.

You should also check to make certain that the queen has enough room for laying. If you find that most of the combs in the brood nest are filled with brood, pollen and honey, you should provide more room. This can be done by adding a super or by replacing combs of honey with empty combs or with frames containing foundation. The combs of honey may be placed in the food chamber super, if you are using a deep super for the food chamber.

LOOK FOR QUEEN CELLS

If you have an overwintered colony, you should check for queen cells during late spring and early summer. Queen cells are usually built

Blow two or three puffs of smoke into entrance.

Remove outer cover, blow smoke in hole in inner cover.

Pry up corner of inner cover, blow in smoke.

Remove inner cover, blow smoke across frames.

18

Place inner cover in front of hive.

Remove all supers above the brood chamber.

Place them on cover behind hive.

Direct smoke across brood chamber frames.

Using hive tool, loosen frame nearest you.

Lift it out, inspect it, lean it against hive.

Now you have room to loosen and inspect others.

Hold frames over hive as you inspect them.

along the bottom edge of brood combs. They are shaped like peanut shells and hang down from the comb. The presence of queen cells generally indicates preparation for swarming. A quick check for the presence of queen cells may be made by tipping back the second story super and looking along the bottom edge of the exposed combs. If you find queen cells containing eggs or larvae, you should take measures to prevent swarming.

After you finish examining a comb, place it back in the hive in the same order as it was before. Combs in the brood nest may have the lower front corner of the comb chewed away by the bees. Make sure that the combs are replaced with that corner toward the front. Otherwise, the bees will chew away the other corner and then both corners will be missing. That will decrease the amount of space available for brood.

After you have finished examining the combs, crowd together those that have been replaced so you will have room for the first comb removed. Then, replace that comb. With your hive tool, crowd all the frames against the farther side of the hive. Then, crowd them back just far enough to equalize the space on both sides of the hive. Do this so there won't be extra space left between any of the frames. Otherwise, the bees will build comb in this space and next time you will crush bees when you crowd the frames together.

While working with the bees, blow a little more smoke on them whenever they start flying off the tops of the frames and threatening you. Try to use as little smoke as possible though. Experience will teach you how much smoke is necessary. This varies according to the season and whether or not nectar is being gathered.

After examining the brood chamber, you may replace the food chamber and examine that, if necessary. Before replacing it (or any other supers), blow a little smoke across the tops of the frames of the brood chamber to drive the bees down out of the way. Then there will be fewer bees crushed as you replace the food chamber or other supers.

When you have completed your examination of the hive, replace the supers, the inner cover and finally the cover.

It is well to weigh down the cover with a brick or a flat stone to prevent its being blown off by the wind.

It is best for a beginner to check his hives once a week during the spring and summer. Don't worry about doing harm to the colony by inspecting them this often. More harm is done to bees by not checking them often enough. It doesn't take long for bees to starve or to make preparations for swarming. If you don't check your bees regularly (about once a week) you can easily lose them and thus a good part of your investment. You should keep up-to-date on conditions within your hives so you can take proper steps to remedy things that go wrong. Neglect of their bees is probably the greatest reason for failure of beginning beekeepers.

HOW TO DEAL WITH STINGS

The sting is a weapon for defense of the colony. In order to reduce the likelihood of its being used on you, keep the following principles in mind:

Try to work the bees when they are flying actively in favorable weather.

Wear protective clothing: A veil over your head and face, gloves for your hands (which you will quickly discard with experience); and closewoven, light-colored clothing sealed at the ankles and wrists.

Always use a smoker when working with the bees.

As you remove the hive cover or a super, apply smoke gently to the exposed bees.

Don't oversmoke—use just enough—this will come with experience.

If you are stung, remove the stinger immediately by scraping it off with your fingernail or any straight edged instrument. Do not try to pull it out, because this will force more venom into your skin.

Since the stinger is barbed, rapid removal can greatly reduce the effect of the sting.

Most beekeepers eventually develop immunity to stings after a few seasons. However, if you become allergic to bee stings, consult an allergy specialist before you become committed to beekeeping.

Beekeeping for Beginners (USDA)

CHAPTER 3

Installing Package Bees

by Enoch H. Tompkins

If you can finance it, it is better to start with two colonies of bees rather than one. If you have two, you can compare their progress and have a better idea if something is wrong with one of them. Also, if something happens to one of the queens you can take a frame of brood with eggs from the other hive and give it to the queenless hive. Then the bees can rear a new queen to replace the lost one. Combs of sealed brood may be taken from the stronger colony and given to the weaker to increase the number of bees in the weaker colony. Combs of honey may be taken from one colony and given to the other when food stores become short. If one colony becomes hopelessly weak in number of bees, it may be united with the stronger hive, then divided later, when the number of bees warrants such action.

Receiving the Bees

A few days before you expect your package bees to arrive, notify your postman that you are expecting them. Give him your phone number and ask him to notify you when they arrive. Generally, postmen are only too glad to do this since they are anxious to part company with the bees as soon as possible. Letting the postman know that you are expecting the bees may prevent their being held at the post office over a weekend. The sooner you can get them the better the condition they are likely to be in.

When you pick up the bees (or when they are delivered to you), you should first check on their condition. It is normal for the package to contain a few dead bees. This is allowed for by the shipper who adds extra bees above the weight ordered. However, if one-fourth or more of the bees are dead, you should ask the postmaster to write this fact on his official stationery for you. Send this statement to the person who supplied your bees and ask for either a replacement or a refund for the portion that was dead. Do not refuse to accept the bees since such action will only complicate matters.

FEED SUGAR SYRUP

When you get the bees home, you should feed them some sugar syrup. The syrup is made by mixing granulated sugar with an equal amount of hot water and stirring it until the sugar is dissolved. The syrup may be made by mixing five pounds of sugar with two quarts of water. Bring the water to the boiling point, then remove it from the heat and slowly stir in the sugar. Don't let the syrup boil or it will caramelize and may cause dysentery when fed to the bees. Let the syrup cool before feeding it to the bees. You should provide at least ten pounds of

Paint brush used to coat package with syrup.

sugar for each package of bees. To feed the bees in the package, you may brush the syrup onto the sides of the screen with a clean paint brush or a piece of cotton cloth, or sprinkle it onto the bees. Lay the package on its side while doing this. Place a shallow pan under the package to catch any syrup that may drip. Feed the bees all the syrup they will readily clean up. This is likely to be about a pint per package. If you are unable to put the bees in a hive right away, they should be placed in a cool, dark, well-ventilated place where the temperature is between 50° and 60° F., such as in a corner of your basement. They may be kept for two or three days, if necessary, but the sooner they are put into a hive, the better. Feed the bees twice a day while they are in storage.

IS THE HIVE READY?

Before your bees arrive, you should have gotten the hive and put it together and painted it. It should be placed in the location where you want the bees to be kept. This location should preferably be in a place that is sheltered from the prevailing wind either by a row of shrubs or trees, a building or a wooden fence (perhaps a snow fence). Try to find a place where the bees won't be noticed by your neighbors. This will prevent them from getting nervous about the presence of the bees and the possibility of getting stung. Don't put them near an outdoor clothesline since they are likely to soil the clothes when they take their spring cleansing flights. Try to keep them away from places where

Shrubs and trees are good wind breaks.   *Snow fence can be used as wind break.*

people will be passing by, unless there is a row of shrubs, trees or a fence to cause the bees to fly up and over the heads of passersby. In the northern states it is best to locate the hives in a sunny place where they will be exposed to sunshine most of the day. This will get the bees warmed up early in the morning and out gathering nectar and pollen. In southern states, a shady location is better where they will at least be protected from the afternoon sun. Such a location will assist the bees in keeping the hive cool in the summertime. Face the hive entrance toward the south or east, again so the sun will shine into the entrance and stimulate the bees to start working early in the morning. Urban dwellers may find it convenient to place their hives on a rooftop to prevent trouble with neighbors.

The hive should be prepared for receiving the bees by removing five of the brood frames. These will be replaced later. The space is needed for receiving the bees from the cage.

INSTALL IN THE EVENING

The bees should be installed in the hive toward evening so they won't be flying much before they have a chance to get settled in the hive. If you are installing more than one package, there is danger of the bees drifting from one hive to another. This possibility is reduced by installing them later in the day. Bees may be installed even when the weather is cool or rainy.

There are several methods of installing package bees in a hive. The method which is thought to be best for the beginning beekeeper is as follows:

When you are ready to install your bees, get your hive tool and bee brush, fill your feeder with syrup and light your smoker. Put on your veil and place the shipping cage near the hive. Lay the package on its side and feed the bees all the syrup that they will clean up readily, by brushing or smearing it on the sides of the cage or by sprinkling it onto the bees. Give the bees time to clean up the syrup. You may wish to place a pan under the cage to catch the syrup that drips through the cage.

The shipping cage comes equipped with a can of syrup for feeding the bees while en route from the shipper. The queen bee comes in a queen cage which is suspended from the top of the bee shipping cage and hangs down among the bees beside the syrup can. Place the shipping cage beside the hive and remove the covers from the hive. Jolt the cage upon the ground to knock the bees to the bottom of the cage. With your hive tool, pry off the cover of the cage and keep it

Sprinkling syrup onto side of bee package.

Use hive tool to pry cover off package.

handy for replacing when needed. Again jolt the bees to the bottom of the cage, then remove the queen cage. This will leave a hole in the cage. Close this hole with the inverted cage cover or a wad of paper or cloth.

Examine the queen cage to see whether the queen is alive. You may distinguish her from any accompanying workers by the fact that she is longer than the workers and her back is smooth and shiny whereas the backs of the workers are covered by hair. If the queen cage has white candy in one of the end compartments, remove the cork, cardboard or tin from the end of the cage in which the candy is located. With a wooden match or nail of about the same size, poke a hole through the candy so the bees can chew it away and let the queen out. Be careful not to injure the queen.

Place the queen cage between the top bars of two of the frames in the hive near the open space left for the package of bees. Suspend the cage, candy end up, by pushing the frames against the cage. The candy end should be up so any dead worker bees won't plug the hole and prevent the queen from emerging.

Poke hole through candy in queen cage.

REMOVE SYRUP CAN

Now jolt the bees to the bottom of the cage as before and remove the syrup can. This may be done by prying it out with your hive tool until you can grab it with your fingers. Or, you can hold the cage upside down and let the can slide out until you can grasp it. Then turn the cage right side up and remove the can. Set the can to one side after shaking or brushing the bees off into the hive. Dump about a pint of the bees onto the queen cage and the rest into the open space in the hive. Place the empty cage in front of the hive with the open top resting against the hive entrance. Any remaining bees can crawl out through the opening and into the hive.

Some shippers of package bees place the queen in a cage that doesn't contain any candy for food. If your queen cage doesn't have any candy, then you should use the following procedure for introducing the

Pry out syrup can with hive tool.

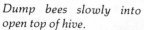

Dump bees slowly into open top of hive.

queen: After dumping the bees into the open space of the hive, wet the queen down with syrup by dripping some through the screen of the queen cage. This will prevent her from flying when you open the cage to release her. Remove the cork or screen from one end of the cage and place your finger over the hole to prevent the queen from escaping. Place the cage on the hive bottom close to the mass of bees that you have dumped into the open space. The queen will come out and join the other bees.

Replace the extra frames and gently push the bees out of the way by rocking the frame back and forth as it is lowered into place. As soon as all of the frames are in place, crowd them to one side of the hive. Then push them back toward the other side of the hive to equalize the extra space between the two sides. The frames should be crowded closely together at all times so the bees won't build extra pieces of comb between the frames. If this happens, it will be messy to remove later and may result in crushed bees when the frames are pushed together at another time.

PUT ON INNER COVER

The inner cover should be put on the top of the hive as soon as possible before the bees boil out over the sides of the hive. If this happens before you can get the inner cover into place, you should brush the bees off the top edge of the hive so you won't crush any when you put the inner cover into place. You may also use your smoker to make them move out of the way by blowing smoke at those on the edge of the hive. Don't worry about those left on the outside of the hive. They will find their way to the entrance and go into the hive later. The entrance feeder may now be placed into the hive entrance and a jar of

Entrance feeder in place.

syrup placed into the feeder block to provide food for the bees while they are building the foundation into comb. The entrance of the hive should be stuffed lightly with green grass to keep the bees confined and protected from robber bees until they can get organized in their new quarters. The grass should be removed the next day and replaced by an entrance block to close the entrance except for a space three-eighths by three inches. The small entrance should preferably be located at the end of the block opposite that of the feeder. That will discourage robber bees from trying to enter the hive to get syrup from the feeder.

A BETTER FEEDING METHOD

A better method of feeding the bees is to place the feeder over the hole in the inner cover. If placed there, it is more accessible to the bees and the syrup will be taken more readily. It will be located in a warmer spot just above the cluster of bees and they won't have as far to go to get to it. Otherwise, the bees have to travel down to the bottom of the hive where it is cooler and out through the cool tunnel to get the cool syrup. On cold days, they won't like to do that. You can adapt your entrance feeder to be used above the inner cover. If the feeder block is made of plastic, just set it above the hole in the inner cover and put the feeder jar in place. Close up the tunnel, that opens to the side, with a wad of paper or cloth or a small block of wood cut to fit the opening.

If the feeder block is made of wood, you will need to cut a hole in the bottom of the block so the bees can come up through the block to get to the syrup. Then plug the tunnel that opens to the side, as suggested for the plastic feeder block.

A friction-top pail may be used for a top feeder by punching half a dozen holes in the center of the cover using one of the small nails that are used for assembling brood frames. A coffee can with a plastic cover may also be used for a feeder by punching the necessary holes in the cover. What you need is something from which the bees may suck the syrup but from which the syrup will not drip. A little may drip out when the container is first inverted, but it should stop shortly as soon as a vacuum is formed. Atmospheric pressure will then prevent further dripping. Keep feeding the syrup until all of the foundation in the brood chamber has been drawn into comb. You will need an empty hive body or two shallow supers to put above the inner cover so you can put the hive cover on and to keep rain out of the hive. A wooden or cardboard box the same size as the hive, or a little smaller, may be substituted for the hive body or shallow supers. Place the telescoping outer cover on top of the empty hive body, supers or box and place a stone or brick on top to keep the cover from being blown off.

Use conventional entrance feeder block.   *Cut circular hole in bottom of block.*

Place block over hole in inner cover.   *Protect jar with hive body, outer cover.*

Do Not Disturb

Don't disturb the hive for at least five days after installing the bees except for feeding. The syrup should be checked every two or three days to see if it needs to be replenished. After about a week, the bees should be checked. Use as little smoke as possible at this time. Take off the outer and inner covers. Remove the queen cage and check to see if she is still inside. If she is, make a larger hole in the candy and replace the cage.

If the queen has left the cage, remove one of the frames that was next to the cage and examine it to see if there are eggs and larvae. If you find these, don't look for the queen but replace the comb, crowd the frames close together and close the hive. If you don't find eggs or larvae on that comb, look on other combs. If you don't find eggs or larvae, chances are that something has happened to the queen and you should immediately order a replacement from your supplier. When she arrives, introduce her in the same manner as the original.

Check Your Bees Each Week

Check your bees weekly, if possible, to make certain that everything is progressing satisfactorily. When the foundation in the brood nest has been all drawn out, you should be able to stop feeding the bees. This is likely to be at the end of the sixth week. It helps to move frames containing undrawn foundation from the sides of the hive and place them between the broodnest and those combs containing honey. Only one frame at a time from each side of the hive should be moved in this fashion. This helps to get the bees to draw out the foundation a little faster. Don't put frames containing only foundation between combs that contain brood. To do so may result in dead brood as the result of being chilled.

When all of the foundation has been drawn out, or when the bees are occupying all ten frames in the brood chamber, it is time to add a super. This first super should preferably be a deep super which will be used for the food chamber to hold honey for the winter stores.

Whenever you see that the bees are having difficulty in getting in and out of the small entrance, you should gradually enlarge the entrance as needed. If they are having trouble getting in and out, they will be clustering near the entrance. The entire entrance space should be opened when the bees have eight or more frames of brood, if the weather has become settled.

CHAPTER 4

Queen Supersedure

by Enoch H. Tompkins

When a queen's egg-laying ability diminishes to a low level, she may be replaced by the workers. This process is called *supersedure*.

Supersedure may be a problem among package bees. Such supersedure is likely to take place about three weeks after the bees are installed in their hive. At that time there exists an imbalance between the number of adult bees and the amount of brood. Many of the adult bees will have died and will not have been replaced by young bees. This leaves the colony with an abundance of brood which is out of proportion to the number of adult bees. The adult bees are all older ones with no bees of intermediate ages. At this time the bees may decide to replace their queen with a new one. Such supersedure may generally be prevented by giving the package bees a comb of emerging brood and adhering bees from a healthy hive about two weeks after they are put into the hive. Such a comb will also help immensely in building up the number of bees in the colony.

Supersedure may also be caused by the beginner working too frequently with the new colony. One shouldn't be tempted to open the hive more than once a week and not sooner than five days from the time package bees are installed.

Poor-quality queens or those infected with the bee disease, *nosema*, may also be superseded. The ovaries of an infected queen soon degenerate and the result may be a queenless colony or supersedure. It is recommended that package bees be fed sugar syrup that has been medicated with the drug *Fumidil-B*, which is available from equipment suppliers. See the later chapter on diseases and pests for more information on this subject.

When bees are preparing to supersede their queen, they will build three or four queen cells upon the face of the comb. Such cells will usually be all of the same stage of development. Swarm cells (built when the colony is planning to swarm) are generally built along the bottom edge of the comb. They are more numerous (ten or more) than supersedure cells, of various stages of development, and generally whiter in color. Sometimes the colony will swarm when the new supersedure queen emerges. At other times the old superseded queen and her daughter will live together peacefully in the same hive for awhile.

Check comb bottoms for queen cells.

CHAPTER 5

Spring Management of Overwintered Colonies

by Enoch H. Tompkins

The purpose of spring management is to get your colonies built up, in number of bees, so the hives are boiling over with bees at the start of your main honeyflow. The following management practices are designed for that purpose. To obtain a large crop of honey you need lots of bees to gather the nectar when it becomes available.

First Spring Inspection

Your first spring inspection of overwintered colonies should be made on a day when the outside temperature gets to about 60° F. This shouldn't be a prolonged inspection because of the danger of chilling some of the brood. Remove the outer and inner covers, as described in the section on working with bees. Check to see if there are combs of sealed honey right next to the cluster of bees. You can check on this by looking between the frames just below the top bars. If you don't see sealed honey adjacent to the cluster, you should take a comb of honey from the side of the hive and place it right next to the cluster. Pry the frames apart where the honey is needed and insert the comb of honey in that space. There is more danger of the bees starving at this time of year than during the winter. When the weather turns cold again, the bees may not be able to reach the necessary honey unless it is placed in close proximity to the cluster. Brood rearing will be in process and plenty of honey is needed for feeding the young bees and for keeping up the temperature of the cluster. If a hive has sealed honeycomb close to the cluster, you should quickly close the hive and move on to the next.

If a hive is short of honey at this time, you may feed it dry sugar, or borrow combs of honey from another hive that has a good supply. Dry granulated sugar may be placed on top of the inner cover in the space provided by the rim. With the hole in the inner cover open, the bees can come up through and eat the sugar. They must dissolve the sugar with water first, so don't feed it to a colony that needs food

Comb of honey, a welcome sight in spring.

Placing dry sugar on top of inner cover.

immediately. Don't feed the bees sugar syrup at this time since that would cause them to rush out of the hive where they might be chilled, on cold days, and not get back to the hive.

Weak colonies should be united with other weak colonies or with strong colonies. They may be divided later if this is thought desirable.

Any hives in which the bees have died, should be closed up tightly to keep out robber bees. These hives should be removed from the apiary as soon as possible to prevent robbing and the possible spread of disease. They should be cleaned of dead bees, inspected for disease and stored in a bee-tight building until they are needed later. An un-heated building is best since there is less chance of the wax moth destroying the combs.

Second Spring Inspection

A more thorough examination may be made at a later date when the outside temperature is around 70° F. This is likely to be at the start of fruit and/or dandelion bloom. At that time you should look for the queen, or at least for eggs and brood. Weak colonies may have no brood at this time, but those covering six or more combs should have. Check the brood for signs of disease (see a later chapter on this subject). Check for drones being reared in worker cells. This could indicate a failing queen which is laying unfertilized eggs. Such a hive should have a new queen introduced or be united with another colony, after first destroying the drone-laying queen.

The brood pattern should be checked to see if the queen is doing a good job of egg laying. Look at a comb of sealed brood. There should be very few open cells in this area. A good queen should lay eggs in most every cell and most of these eggs should develop into adult bees. If there are many scattered, open cells in the sealed-brood area, this may indicate a failing queen which should be replaced.

UNITING COLONIES

Colonies may be united by removing the covers from one hive, placing a sheet of newspaper on top and placing the other hive (with its bottom board removed) above the newspaper. One should first punch about a dozen small holes in the paper with a common pin or a toothpick, so the bees can get hold of it to chew it away. While the bees are chewing the paper, they become acquainted, and gradually the odors of the two hives intermingle into one. They are thus able to unite peacefully with no loss of bees. The queens will fight and the more vigorous one will survive.

SPRING FEEDING

On the second examination, one should check more thoroughly on the amount of honey in the hive. If there is less than the equivalent of three brood combs full of honey in a hive, the bees should be fed.

Combs of honey, from other hives that have a surplus and are known to be free of disease, may be given to those colonies that are short of stores. Or, one may feed sugar syrup made from equal parts (by volume or by weight) of granulated sugar and hot water. The water should be brought to the boiling point then removed from the fire. The sugar is then stirred in until it is dissolved. The syrup should not be boiled because that might scorch it and make it unfit for bee feeding. The syrup should be allowed to cool to lukewarm before feeding it to the bees.

UNITING COLONIES

A simple way to unite colonies is to take covers off one hive, place a sheet of newspaper on top (top photo), and punch several holes in paper with a pin or toothpick. Then, as in lower left photo, place second hive, with bottom removed, on top of paper. Left for a few days, as at lower right, bees will chew away paper and unite without difficulty.

There are a number of types of bee feeders on the market. The Boardman entrance feeder is the one that is commonly supplied with beginner beekeeping kits. This feeder is more efficient if placed over the hole in the inner cover as explained in the chapter on installing package bees.

A simple feeder may be made from friction-top pails, coffee cans with plastic covers, screw-top canning jars, used mayonnaise jars or seven-pound plastic honey pails. All that is needed is half a dozen nail holes punched in the center of the cover to provide openings for the bees to suck out the syrup. Use a nail the size of those used for nailing brood frames together. The holes should be punched from the inside out so the bees won't damage their tongues on the burr around the edge of the holes. These feeders are placed above the hole in the inner cover where the bees may feed in comfort without extra traveling to an entrance feeder. The feeder should be checked twice a week to make certain that it doesn't run dry. Test your feeder to make certain the syrup doesn't continue to drip out. When you turn it upside down, some syrup will drip until a vacuum is formed, then it should stop.

Three types of feeders are shown here. In rear is a division-board feeder, put in a hive to replace a brood frame. At left is a friction-top can, and at right is a Boardman entrance feeder.

DIVISION-BOARD FEEDER

Another type of feeder in common use is the *division-board feeder*. This is shaped like a brood frame. A brood frame is removed from one side of the hive and the feeder is put in its place. Syrup is poured into the top by moving the inner cover a little to one side. The bees crawl down through the top to get the syrup. A thin piece of board should be floated on top of the syrup for the bees to stand on so they won't drown. Or, a piece of aluminum or plastic window screen may be shoved down into the feeder so it provides a runway down each side of the feeder. Some division-board feeders are made of wood and some are made of plastic.

Feed whatever amount of syrup you think is needed to keep the stores above the minimum of fifteen pounds of honey or syrup. It is better to feed more than is needed than not enough. If the bees don't use it right away, they will have it for later. It won't be wasted. Stop feeding at the start of your main honeyflow so you won't be contaminating your honey crop with sugar syrup.

When feeding your bees, it would be well to add medication to the syrup as a method of preventing American Foulbrood infection and for combatting nosema. See the section on diseases for further information.

REMOVE WINTER PACKING

If you have provided winter wrapping for your colonies, this may be removed at the time of your second spring inspection.

REVERSE HIVE BODIES

During the winter and early spring, the bees tend to move their cluster into the upper stories of the hive. During the second inspection, it is well to reverse the position of the food chamber and the brood chamber of strong colonies (those covering ten or more combs). This will relieve congestion and give the queen more room for laying. This is a good time to clean the debris off the bottom board. This may be speeded up by having an extra bottom board with you which may be substituted for the original bottom board, which may then be cleaned and made ready for the next hive. Make certain that the entrance block is replaced at this time.

In spring, when outside temperatures reach 70° F., beekeepers reverse the position of the two hive chambers, moving brood chamber (#1 in picture) to top, and placing food chamber (#2) on bottom.

FIND THE QUEEN

During fruit bloom is a good time to find your queens. There are fewer bees in the hive now and the queen will be easier to find. The queen is longer than the worker bees, and is generally surrounded by a circle of attendant workers. She is most likely found on a comb that contains eggs. The back of her thorax (middle section of her body) is shiny and bald as compared with the hairy thorax of the workers and drones.

Robbing by Bees

During times when there is no nectar available from flowers, bees may start robbing honey from other colonies. Usually those being robbed are colonies that are weak in number of bees and are unable to defend their hives. One should always see that such hives have their entrance reduced to a size that can be well guarded.

Robber bees may be distinguished by their furtive, darting motions around hives. They are likely to be exploring the cracks between supers. Guard bees pounce upon them when they try to get into the entrance of hives.

It is better not to work with your bees when there is no nectar being brought into the hive, if this is possible. If you do have to work with the bees at such a time, you should provide an empty hive body with bottom board (having a closed entrance) and a cover. Any combs that must be removed from a hive should be placed in this empty hive and quickly covered so robbing won't be encouraged. Never leave any combs of honey or pieces of honeycomb exposed to bees at such a time. To do so may start robbing. Robbing quickly spreads throughout the whole apiary and upsets the normal routine of the bees. It results in many bees being killed, either as robbers or in defense of their hive.

If robbing should start, you should reduce the entrances of all of your hives to a size that may be easily defended, according to the strength of the individual hive. For weak hives, an entrance about three-eighths by one inch would be suitable. Placing a few handfuls of grass against the entrances will discourage robbers and help with the defense of the hive.

Requeening

When a failing queen needs to be replaced, or when something happens to an original queen, replacement queens may be purchased from southern bee breeders. The names of these suppliers may be obtained from one of the beekeeping magazines mentioned in a later section of this book. Most beekeeping equipment suppliers also have queens to sell. The replacement queen should be ordered and received before making preparations for installing her in the hive. They are shipped by airmail in a wooden cage having three interconnecting cavities and covered by a piece of wire screen. One of the end cavities is filled with a candy which provides the queen and a few worker attendants with food for the trip. When the queen arrives, give the bees a drop of water on the wire screen. If you have to keep her in the cage a few days, give them a drop of water twice a day.

When you are ready to put the queen into a hive, the worker bees should be removed from the cage. The bees in the hive will likely accept the queen, after they have gotten acquainted with her, but may be antagonistic toward the workers. The workers may be removed by taking the cage into a small room with a window. A bathroom is good for this purpose. Don't turn on any lights, but take the cage to the window. Remove the cork or piece of screen which covers the end of the cage that is opposite the candy. Hold the cage against the window

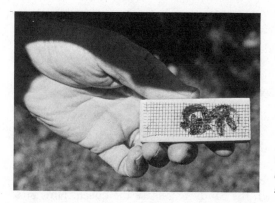

Queen and several workers are packed into tiny wooden shipping cage.

pane with the screened side against the pane. Tap the cage with your finger and one at a time, the bees will come out onto the glass. The light attracts them so they aren't likely to stray far from the window. They will buzz about but aren't likely to sting. When all of the workers have left the cage, replace the cork or screen, if the queen is still in the cage. If she also came out, you will have to pick her up by both wings and place her head into the hole in the end of the cage. Don't pick her up by her abdomen or you may injure her. If you don't wish to pick her up, you may place the cage against the window pane and herd her into it with your fingers. When you get the queen into the cage, replace the cork or screen over the hole. Then you may kill the worker bees. Or, you may wish to kill them as they come out of the cage. Make sure that you don't kill the queen by mistake.

DISPOSE OF OLD QUEEN

Before you can requeen a colony of bees, you must first find and dispose of the old queen. Following the general procedure for working with bees, you should remove the first or second comb from the brood chamber. Look this comb over carefully to see if the queen is on it. If not, then remove the center comb and see if she is on that. If you don't find her there, continue to check the combs adjacent to the center comb and work toward both sides until you find her. Always replace the combs in the hive in the same order as they were removed. If you don't find the queen on the first round of inspection, close the hive for awhile to let the bees resume their normal activity, then try again. When you find the old queen, you may dispose of her as you see fit. You may kill her by squeezing her thorax between your thumb and forefinger. While you are looking for the old queen, you should also destroy any queen cells that may be present. If any are left, the colony won't accept the new queen. If the colony is queenless, the queen cells should also be found and destroyed before introducing the new queen.

INTRODUCING THE QUEEN

After the above preparations, you are ready to introduce the new queen. This is done by first removing the cork, or screen, or cardboard, which covers the hole in the candy end of the cage. This exposes the candy. Next, you should poke a hole through the candy with a wooden match, or a nail of about the same size. That will make it easier for the bees to chew a hole through the candy to release the queen. Now suspend the cage between two frames of brood (if present) with the candy end up. The candy end should be up to prevent any dead bee remaining in the cage from plugging the exit hole and preventing the queen from getting out. The hive should then be closed and left undisturbed for about a week. The bees will chew a hole through the candy and release the queen. After a week, the hive may be opened and the cage checked to see if the queen got out safely. If she should still be in the cage, make a hole in the candy, large enough for the queen to get through and replace the cage in the hive. If she isn't in the cage, remove it and check the adjacent combs for eggs. The presence of eggs indicates that the queen has been successfully introduced. If you can't find eggs, or the queen, you will have to introduce another queen as soon as possible. It may be difficult for a beginner to find either eggs or the queen. If you can't, you may want to call in an experienced beekeeper to make a double check for the presence of the queen.

PICK THE RIGHT TIME

Queens are more safely introduced when nectar is being brought into the hive. The best time for this operation is during the fall honeyflow. Most beekeepers requeen at this time so the new queen will have time to produce plenty of young bees before winter. If you have to requeen a colony when there is no nectar available, feed the bees on sugar syrup for a few days before introducing the new queen and for a few days afterward. That will simulate a honeyflow.

LAYING WORKERS

When something happens to a queen and the bees are unable to replace her by rearing a new queen, worker bees may develop the ability to lay eggs. Perhaps the colony reared a new queen but she may have gotten lost or eaten by a bird while on a mating flight. Worker bees are sexually undeveloped females and the absence of brood in the hive allows some of them to develop to the extent that they can lay eggs. Worker bees are unable to mate so they can lay only unfertilized eggs. These hatch into drones so the colony is no better off than before. Laying workers are likely to develop about six weeks after the old queen disappears. At that time there is no longer any brood in the hive which would have an inhibitory effect upon the development of laying workers.

A case of laying workers may be determined by the presence of more than one egg in a cell. There may be from two to perhaps a dozen in a single cell. The cells containing eggs will be scattered about the comb in a random manner rather than in a uniform pattern as when deposited by a queen. Some of the eggs will also be deposited upon the sides of the cells. Since the eggs are unfertilized, they will develop into drones. When these cells are capped, they may be recognized by the fact that they protrude from the face of the comb like the rounded nose of a bullet. Since the drones will have been reared in worker-sized cells, the adults will be smaller than normal. After laying workers have been operating for an extended period, there will be an abundance of drones, as compared to the number of workers.

It is probably impossible to replace laying workers with a new queen. The other workers regard the laying workers as they would a queen and would kill a new queen if she were introduced. It's next to impossible to distinguish the laying workers from the others so as to eliminate them from the hive. The best solution to a colony having laying workers is to unite it with a strong colony, using the newspaper method described elsewhere in this book. The hives may be divided again, later, if this seems desirable.

Swarming

Swarming is the natural method by which bees increase the number of colonies. It generally takes place in late spring or early summer. Bees are likely to swarm during the light honeyflow that precedes the major honeyflow. Overcrowding or a failing queen contribute to the desire to swarm. Bees may also swarm when a heavy honeyflow is broken by alternate periods of rainy and fair weather.

Before they swarm, the bees will build a number of queen cells. Queen cell cups are likely to be present at any time of the year. But, when the bees are preparing to swarm, they start building these cups out and enlarging them. The new part of the cell is likely to be made of lighter colored wax than the cups, which will be the same color as the comb upon which they are built. When you see these cells being enlarged you will know that preparations for swarming are under way. Another sign of swarming is when the bees start to cluster on the landing board of the hive and around the entrance. However, this clustering may take place during warm weather when the bees are trying to cool the hive. Or, it may take place at the end of a honeyflow when the bees have nothing to do.

Bees are likely to swarm at about the time that the first queen cell is capped over. The swarm is most likely to issue between 10 A.M. and 2 P.M. About half of the bees leave the hive, generally with the old queen. They are likely to cluster on the branch of a nearby tree while they are waiting for scout bees to locate a new home for them.

HIVING THE SWARM

The swarm should be placed in a hive as soon as possible. They may stay in their original location just a few minutes or for a few days. If the bees cluster on a low branch, they may be hived quite easily. Place a sheet of plastic or other material on the ground under the swarm and place the empty hive on it, with the entrance to one side of the swarm. If the swarm is only three feet, or less, off the ground, they may be shaken off the branch and onto the sheet in front of the hive. They will then run into the hive of their own accord. If they are clustered higher, you may have to cut off the branch and lay it in front of the hive or shake the bees off in front of the hive. Swarms that cluster high in a tree are best ignored unless they may be easily shaken into a box or basket and lowered to the ground.

Sometimes the swarm may cluster on the side of a tree, a fence post or on the wall of a building. In such a case they may be scooped off with a shoe box, or similar container and dumped in front of, or into, a hive. Or, you may be able to place the hive on the ground or on a box with the hive entrance touching the swarm. They will generally be glad to go into the hive. Another way is to gently brush them off in front of the hive or into a box from which they may be transferred to the hive.

If an empty hive is not available, the bees may be temporarily housed in a wooden or cardboard box. A hole or slot should be provided for an entrance.

When most of the bees are in the hive, it should be moved to its permanent location.

Swarm of bees dumped on sheet in front of hive, as in upper picture, will soon begin to enter hive, as in lower picture.

SWARM PREVENTION

You will get more honey from your bees if you can prevent swarming. Swarming divides the efforts of the colony and removes most of the nectar-gathering force at the time it is most needed. There are several things that the beekeeper may do to cut down on the possibility of swarming:

1. Provide plenty of storage room for incoming nectar and surplus honey when nectar becomes available. Keep ahead of the needs of your bees in this respect. Add supers a little before they are needed rather than afterward. Nectar may contain 80 percent water when first brought into the hive. Honey, the end product, contains only about 17 percent water. The bees need room to store and to process the nectar. Make certain that your bees have enough room to do the job properly. Probably the greatest failing of beginners is not providing enough super space at the right time. Don't wait until one super is filled with honey and capped before adding another super. When you see that the bees are working on eight frames of your top super, give them another one. If they don't need it, they won't use it. It is better to have it and not need it than to need it and not have it. Adding supers not only gives the bees room to store the nectar and honey but it also gives the bees room to cluster when they are not working. This cuts down on hive congestion, one of the causes of swarming.

2. Bees are less likely to swarm when headed by a queen that is less than two years old. Such a queen secretes a larger quantity of glandular substances than an older queen. These substances (pheromones or queen substance) are passed around among the worker bees and inhibit the production of supersedure queen cells. Requeening every year or two helps to control the swarming impulse.

3. Provide the queen with sufficient space for laying eggs. Make sure that the brood chamber isn't getting decreased by too much honey being stored in it. The queen should have empty cells available for egg laying at all times. You may have to remove some combs of honey from the sides of the brood chamber and replace these with empty combs or frames of foundation. Empty combs are best. The frames of foundation are best reserved for supers and for a time when nectar is coming into the hive. The combs of honey may be placed in the food chamber or given to a colony that needs more honey. Most, if not all,

of the combs of the brood chamber should be reserved for brood rearing.

During the winter and early spring, the bees tend to move upward in the hive and bring their brood nest with them. In a two-story hive, additional space for brood rearing may be provided by periodically reversing these two parts of the hive during the spring buildup preceding the main honeyflow. This practice will relieve congestion of the brood nest and help prevent swarming.

4. When a hive becomes congested with bees, and the weather becomes hot, bees may have a problem in keeping the hive cool enough. When you see bees clustering outside the hive entrance on hot days, you should do something to assist them in cooling their hive. This may be done in several ways.

One method is to give the hive a larger entrance. If the entrance block is still being used, change it to the larger-sized entrance. Or, you may wish to remove the entrance block and substitute one that closes only half the entrance. If the entire entrance is open, you may provide more air circulation by propping the front of the hive body up with small blocks of wood at the two front corners.

Another method of providing additional ventilation is to lift up the outer cover of the hive and move it forward so the bottom rear edge rests upon the edge of the inner cover. With the hole in the inner cover open, top ventilation is provided around the edges of the outer cover.

Ventilation may also be provided by moving the inner cover toward the front of the hive about half of an inch with the outer cover in the position just described. Or, the top supers may be staggered to provide half-inch spaces at front and rear.

In southern states, it is well to place your hives where they will be shaded during the hottest part of the day. Or, you may want to make

Two ways to ventilate hives are, left, to lift and push back outer cover so one edge rests on inner cover, and, right, to stagger the supers to provide half inch spaces in front and rear.

a slatted rack of lath to place on top of the hive to provide shade. Such a rack should be weighted down by bricks or a stone so it won't blow away.

5. Provide an adequate supply of water for the bees if there is not a source of fresh running water nearby. Water is used, at times, to dilute the honey for feeding brood. It is also brought into the hive and placed on top of frames and in burr comb to help cool the hive through evaporation. Fresh running water is preferable. This may be provided by letting an outside faucet drip onto a slanting board. The bees can help themselves to it as it runs down the board. If you would rather have the bees farther from the house, a hose may be attached to the faucet and the water allowed to drip onto a board at the other end of the hose. Stagnant water, as a source for bees, is thought to assist in the spread of the bee disease, nosema.

6. Combs of sealed brood may be removed from the brood chamber and empty combs, or frames with foundation, may be put in their place. Place these beside other combs, containing brood rather than break up the brood nest by placing them between brood combs. The combs of sealed brood may be placed in a super above other supers on the same hive, or may be given to weaker hives to strengthen them. Removing sealed brood relieves congestion in the brood nest and gives the queen more room for laying eggs.

7. Eliminate combs that have large areas of drone comb. Drones are large bees that take up room and get in the way of the workers. Only a few drones are needed for mating with your queens. They also eat honey that could be contributing to your honey crop. By removing combs that have a lot of drone comb, and substituting worker combs or frames with brood foundation, you cut down on the number of drones and increase the number of worker bees for storing honey. This also helps to relieve congestion in the hive and helps prevent swarming.

SWARM CONTROL

If the bees start building queen cells in spite of your efforts at prevention, then other measures are called for. One of these is to cut out or destroy the queen cells every eight to ten days. They may be destroyed by sticking the point of your hive tool into them in such a manner that the larvae are killed. You must be very thorough in destroying queen cells since if only one is missed the colony is likely to swarm. Before destroying any queen cells, you should check them closely to make certain they are not supersedure cells. If you were to destroy all such cells, your colony might be left queenless (see the section on supersedure for information on distinguishing one type of queen cell from the other). If the cells are supersedure cells, leave one for the new queen to emerge, and destroy the remainder. Also destroy the old queen, if you can find her.

If destroying queen cells doesn't stop the desire to swarm, you may want to divide the colony and increase your number. It would be better to do that than to risk losing the swarm if it should come out on a day when no one is at home. Or, they might swarm unnoticed by you or others who may be at home at the time. The division may be reunited with the parent hive at a later date if that seems desirable.

There are several methods of dividing a colony of bees. Probably the simplest is to remove four to six combs of brood, with adhering bees and old queen, to an empty hive. Fill the remaining space of both hives with empty combs or frames with brood foundation. Move the new division, with the queen, to a new location. This may be near the old location, but preferably with the entrance turned at right angle to the parent hive. The parent hive should be given a ripe (capped) queen cell or a new, young queen within two hours, or no later than twenty-four hours, after making the division. Dividing should be done during the middle part of the day when many of the bees are away from the hive.

CHAPTER 6

Summer Management

by Enoch H. Tompkins

The main concerns during the summer are swarm prevention, supering for honey production, removing the honey crop, and processing and packing the honey. If you have more honey than you want for your own use, you may be concerned with marketing some of the surplus honey. Swarm prevention and control, and hiving swarms have already been covered in a previous chapter. We will start our discussion of summer management with supering for honey production.

Supering for Bulk Comb Honey Production

It is recommended that the beginner beekeeper produce bulk comb honey, at the start, since there is less investment in equipment needed than with extracted honey production. The production of section comb honey requires a lot of skill and is better reserved for the more experienced beekeeper.

Shallow supers ($5^{11}/_{16}$ in. deep) are most used for producing chunk or cut-comb honey. They may also be used later for extracted honey production. They are lighter and easier to handle than the full-depth supers. A shallow super will hold about 30 pounds of honey compared with about 50 pounds for a deep super.

There are several types of frames that may be used in these shallow supers. I would recommend the type with the thin top bar that has a slot in it for inserting the foundation. A divided bottom bar will allow the sheet of foundation to extend through it and help hold it in place in the middle of the frame. The thin top bar takes up less space than the thick top bar and leaves more room for storing honey in the frame.

The size of frames in shallow and full-size supers are compared. Top photo is frame from a shallow super. Both have new foundation, which has not yet been drawn into cells by the bees.

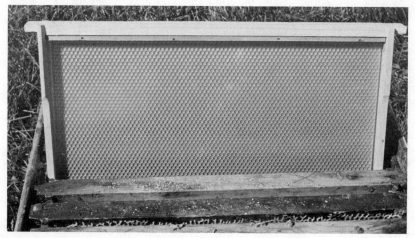

It also allows a deeper comb which more nearly fills the plastic boxes that are commonly used for cut-comb honey.

Three types of foundation are available for comb honey production. For cut-comb and chunk honey it is better to use the special foundation designed for this purpose. The thin super foundation, made for use in comb honey section boxes, is really too thin to be used in the shallow frames. It is more likely to sag and buckle and produce wavy combs. The foundation made especially for chunk or cut-comb honey is a little thicker and will remain straighter in the frames. A new type of foundation for chunk honey has been invented which is supposed to discourage the queen from using it for brood rearing. It is called the "special 7-11 cut-comb foundation." It has cells embossed in the wax which are larger than worker cells but smaller than drone cells. With this type of foundation, you are supposed to be able to eliminate the queen excluder. The manufacturer claims that the queen won't lay eggs in cells of this size.

Sheets of foundation for the split-top bar and divided-bottom bar frames should measure 5¼ x 16¾ inches.

Supering of strong overwintered colonies may start with the blooming of dandelions and fruit trees. Supers will be needed at least by the time clovers start to bloom. For newly installed package bees, the first surplus honey super should be added above the food chamber when the foundation in that super has been drawn out into comb and bees are covering all frames.

USE OF QUEEN EXCLUDER

Place the new super above the food chamber and, when the bees have started building out the comb, check these combs to see if the queen has moved up onto them. If she has, gently brush her off with your finger so she will fall upon the frames of the food chamber. Then place a queen excluder on top of the food chamber and replace the new super. Use of the queen excluder is advised for keeping the queen from laying eggs in your comb honey supers. Any brood that is reared there will turn the comb brown and spoil the appearance.

When the bees are covering eight frames in the first surplus honey super, it is time to add the second super. Remove the first super and add the second right above the food chamber. Place the first super above the second super. This method, called "bottom supering," will get the bees to work on drawing out the foundation faster than if the second super is added on top of the first. It also helps relieve congestion in the brood nest by getting the comb builders up into the super

SUPERING

For bulk comb honey production, these are the steps for supering. At left above, the first shallow super (#3) is placed on the brood chamber (#1) and the food chamber (#2). Second super (#4) in photo above right, is added under the first one. Third super (#5) is placed under second (#4) in photo at right.

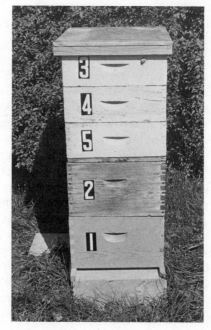

and out of the way of the nurse bees. Another advantage is that the nearly finished honeycombs will be at the top of the hive where they may be removed when they are completely capped.

When the bees are covering eight frames in the second super it is time to add the third. The first two supers are removed and the third is added just above the food chamber. The second super added is placed above this third super and the first super is placed on top. The bees will continue to work on the partially filled supers and will cap the cells when the honey is ripe. Additional supers should be added in the same manner until the honeyflow has reached its peak. From then on, new supers should be added on top of the hive so you won't end up with a lot of partially-filled combs.

Bees have a tendency to start work on the foundation in the middle of the super before working on that next to the sides. When the foundation in most of the frames has been drawn into comb, you can get the bees to work on the outside frames faster by moving them into the middle of the super.

Supering for Extracted Honey Production

Supering for extracted honey production is done in a similar manner as that for chunk comb honey. There are a few differences, however. One difference is that most beekeepers don't use a queen excluder for extracted honey. If the queen lays a few eggs in the extracting combs, this doesn't cause any great difficulty. She will generally be crowded back into the brood nest by the honey being stored in the combs of the super. As soon as a super is filled with honey, it will act as a queen excluder since the queen is reluctant to cross over combs of honey to get to empty combs above.

"Top supering" is used in extracted honey production rather than "bottom supering." In "top supering," additional supers are added on top of those already on the hive rather than directly above the food chamber super as in "bottom supering."

Shallow supers are recommended for extracted honey. They are lighter in weight than deep supers and place less strain on the back when lifting them. A shallow super holds about 30 pounds of honey as compared with about 50 for the deep super. Added to these weights are the extra weight of the frames and of the supers themselves. The shallow supers are the same type as for chunk honey production and use the same type and size of frame. The exception is that heavier foundation

In supering for extracted honey production, additional supers are added on top of those already in place. Thus, in this photo, #3 was added first, #4 second, and #5 last.

is used for extracted honey. The foundation is also reinforced with wires to prevent breakage of the combs while in the extractor. Foundation may be purchased with the wires already embedded into the wax or you can get plain foundation and do your own wiring and embedding. Frame wiring kits and wire embedders are available from suppliers of beekeeping equipment. Some foundation is available that is reinforced by a midrib of plastic, making wiring unnecessary.

USE FEWER FRAMES

In supers for extracted honey, most beekeepers use only nine or eight frames of drawn combs. The reason for this is that the bees build the combs thicker, which makes uncapping easier. Also, costs of equipment are less when fewer frames are used. The frames must be evenly spaced so you don't get some thick combs and some thinner. If too much space is left between combs, the bees may build extra comb in this

When only eight or nine frames are used in a super for extracted honey, the additional space in the super must be proportioned between all of the frames.

space. Ten frames are still used when starting with only foundation in the frames. Otherwise, the bees will build combs between the frames.

By producing extracted honey, the combs can be used over and over again, year after year. This saves the bees much work in building combs and results in larger crops of honey. It also saves on the expense of replacing the foundation every year as is necessary with chunk honey production. However, more time is spent in uncapping, extracting, straining, heating, and bottling the extracted honey. Also, the extracted honey generally sells for a lower price per pound. It is hard to say which type of honey will give the higher returns per dollar invested for the beginner. Extracted honey, no doubt, lends itself to easier handling by the commercial beekeeper who has mechanized equipment.

Removing the Honey Crop

When the combs of surplus honey are all capped over, they are ready for removal from the hive. Use of the bee escape is probably the best method for the beginner to use for removing surplus honey. The bee escape is a metal device that is fitted into the hole in the inner cover. The bees can pass down through the bee escape but are prevented from returning by a pair of flat spring gates. The inner cover, with the bee escape, is placed under the super containing the honey that is to be removed from the hive. The outer cover is placed on top of this super. Make sure there are no holes or cracks through which bees can reenter the super. In 24 hours most of the bees should be removed from the super. If there is some brood in the super, some of the bees will remain there to care for the brood. These may be removed by brushing or shaking them off the combs.

A bee escape is placed in position in hole in inner cover. This cover will be placed under the super containing the honey that is to be removed from the hive. Thus bees in this super can go down into the hive below the inner cover, but cannot return. Most of the bees will find their way out in 24 hours.

Packing Cut-Comb Honey

After the super is removed from the hive, the frames of comb honey may be cut into chunks, four inches square, and packaged for selling. The comb may be placed on a wire cake rack (12½ inches x 18 inches) or on a rack made of galvanized hardware cloth attached to a wooden frame. One-quarter-inch mesh hardware cloth may be attached to a rectangular frame made of pine strip of about ½-inch x ¾-inch dimensions. This is available from lumber companies as "window stop" stock. For the frame for this rack, you will need nine feet of the window stop. You will probably have to buy a ten foot length. Cut this into five pieces 16½ inches long and two pieces 11½ inches long. Using nails the size of those used for nailing frames together (1¼ inches long), nail the short strips (ends) onto two of the long strips (sides) to form a rectangular frame. Nail one more of the long strips in the middle of the frame parallel to the side strips. Nail the other two long strips parallel to the others so they equally divide the remaining space on each side of the middle strip. This completes the cutting frame.

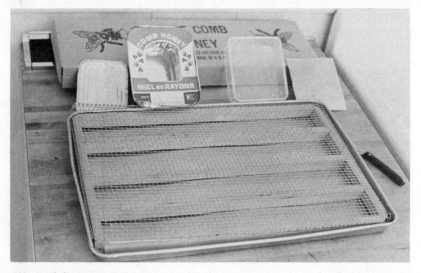

Cut-comb honey packaging needs include, in rear, a package that holds an entire comb. On top of it in center are two sample packages. To left and right of them are guides for cutting the chunks to the proper size. In front are a rack and pan used in cutting the comb.

Ordinary kitchen knife is used to cut comb to proper size, using one of metal squares as a guide.

For covering this frame, you will need a piece of ¼-inch mesh galvanized hardware cloth measuring 12 x 18 inches. Cut a ¼-inch square out of each corner and fold the screen down ¼-inch over the sides and ends of the wooden frame. You now have a rack for cutting two shallow frames of comb honey placed side by side on the rack. This rack will fit into a cookie pan measuring 18 inches x 12 inches x 1 inch that will serve to catch the honey that drips from the cut edges of the comb. An aluminum pan is recommended since it will not rust or turn the honey black as worn tinware would.

A guide for cutting the chunks of honey may be made by cutting a 4-inch square piece of sheet aluminum or piece of a tin can. This guide is laid on top of the comb, starting at one end of the frame, and a paring knife can be used to mark around it to show where to cut the comb. Match the edge of the guide with the inside of the top bar of the frame and the first knife mark, then mark around the guide for the second chunk. Continue in this manner until you have marked out the four chunks. There will be a little bit of the comb remaining at the end and at the bottom of the frame. Don't worry about that. The frame is not generally completely filled with capped comb, so having a little extra allows you to cut chunks that are more nearly all capped over, thus making a nicer looking chunk of honey. It is a good idea to start your first cut about one-half of an inch away from the end bar, for this reason.

Now, use your paring knife, with a sawing motion, to cut the comb along the top bar of the frame, along the ends and along the bottom bar. Lift off the frame and place it in a pan to drain. Next, cut the chunks of comb by following along the marks that you made around the cutting guide. The small strips that are left over may be put into a bowl for home consumption. After you have finished cutting the comb, the pieces should be separated a little to allow the honey to drain away from the cuts. After most of the honey has drained from the chunks, they may be placed into plastic boxes to be sold or for gifts. A short, wide-blade spatula is useful for placing the chunk into the box. These boxes are available from some equipment dealers. They make a nice looking pack and protect the honey from dust, insects and probing fingers.

One Canadian supplier of beekeeping equipment sells aluminum trays for packing cut-comb honey. These take a smaller chunk than the plastic boxes. You can make a guide for cutting chunks for these trays by cutting out the bottom of one of the trays and using that. You will get five chunks from a shallow frame rather than the four for the plastic boxes. The aluminum trays make a nice pack but probably won't stand up well for shipping. The cardboard cover of the tray has a little plastic window and is held in place by folding the sides of the tray over onto the cover. After the sides' are folded over, the corners are folded in to complete the sealing.

Another method of packing cut-comb honey is to cut it into chunks and place these into glass jars. Liquid honey is poured in around the chunks to fill the jar. The liquid honey is first heated to 150° F. then cooled to 110° F. before being added to the jars. This type of pack is commonly called chunk honey. Wide mouth, one-pound jars or square two and one-half and five-pound jars are also used for this type of pack. One chunk is placed in the one-pound jar and two chunks are generally placed in the two and one-half-pound jar. The five-pound jar will take a square chunk on each of the four sides, plus a small chunk in the middle. The liquid honey is likely to crystallize shortly, so don't pack much of it at one time.

Bulk Comb Honey

Bulk comb honey may be sold in the frame. Cardboard cartons are available for packing these. Combs should be wrapped in plastic film before placing them in the carton.

After the honeycomb has been cut from the frames, the frames should be scraped clean, fitted with another sheet of foundation wax and replaced in the super for use on your hives when needed. Replacing the foundation is made easier if the slot in the split top bar is extended to the two ends. This may be done on a bench saw or with a hand saw. If only one-half of the top bar is nailed to the end bars, the other half may be used to wedge against the foundation as that half is put into place. The sheet of foundation should extend out above the top bar about one-eighth of an inch and this part is folded over the top bar to keep the foundation from slipping down.

If you don't completely split the top bar, then you must clean out the slit so you can replace the foundation in the frame.

Extracting Honey from the Combs

Extracted honey is liquid honey that has been removed from the honeycomb. This is generally done by centrifugal force in a machine called a honey extractor. These machines are available in a variety of sizes

A roomy kitchen provides enough space for the entire operation involved in extracting honey. At left, Enoch Tompkins uses electrically heated uncapping knife, with uncapping tub to catch the honey-smeared cappings. In center is a four-frame extractor, which spins the honey out of the cells. At right is a holding tank.

from two-frame to 72 frame. The two-to-four-frame extractors are large enough for the beginner. The three-frame extractor has some advantage in that it may be used for extracting the honey from the cappings.

A four-frame extractor consists of a square basket made of heavy wire mesh that rotates within a metal or plastic drum. The uncapped combs are placed one against each side of the basket. The basket is turned by a hand crank or an electric motor. The centrifugal force, created as the basket rotates, pulls the honey out of the cells. The honey hits the side of the drum and runs to the bottom where it is drained off through a spigot located at the bottom of the drum.

Before the honey can be extracted from the comb, the cells have to be opened. This is done with an uncapping knife which is used to cut a thin slice of wax (the cell cappings) off each side of the comb. Uncapping knives are available in several styles from the plain "cold" knife to the electrically-heated, thermostatically-controlled knife. The latter is recommended as best in the long run.

The "cold" knife is a heavy broad-bladed knife that is generally heated in a pan of boiling water while being used. One should have two of them so one knife can be heating while the other is being used for uncapping the combs.

The thermostatically-controlled electric knife maintains a fairly constant temperature and is ready for use at all times. Also, it doesn't overheat.

Electrically heated uncapping knife is long enough to slice across the entire face of the frame, hot enough to cut easily through wax of cappings.

Place a strip of wood across the uncapping tub for supporting the combs while uncapping them. A piece of pine board two inches wide may be used. It should be about four inches longer than the diameter of the tub. An eight-penny nail (or larger) should be driven through the middle edge of the strip. This should extend about an inch through the strip. At each end of the strip, cut a one-inch wide notch half way through the strip. These notches should fit down over the edges of your tub to help hold the comb support in place. They should be cut on the same edge as that into which the nail was driven. The point of the nail should be pointing upward. A short cross piece of the same material (¾ inch x 2 inches) should be attached to the comb support at right angles and about one-third the distance from one end to the other. A one-inch notch should be cut at the end of the cross piece to fit down over the edge of the tub.

When uncapping, the comb is supported by placing the center of the end bar upon the point of the nail in the comb support. One hand is placed on the other end of the frame to hold it upright over the tub.

A basket made of galvanized or aluminum window screen may be placed in the tub to catch the cappings and allow the honey to drain off. The screen should be supported by a wooden rack about three inches deep to allow space for the honey.

REMOVING THE CAPPINGS

The cappings are removed by starting the knife at the top of the frame, with the edge of the knife resting against both the top bar and the bottom bar. Using these bars as guides, the knife is seesawed sidewise as it is moved from the top end of the comb to the bottom. The comb should be slanted a little toward the knife so the sheet of cappings will fall into the tub below. The frame is then rotated to bring the other side of the comb into position for uncapping. The process is repeated for that side.

As the combs are uncapped, they are placed in the extractor, one against each side of the basket (for a four-frame extractor). When the extractor is full, the basket is rotated by turning the crank or by starting the motor. The machine should be operated at about 200 revolutions per minute for two or three minutes. Only about half of the honey should be removed from the first side of the comb. This is to prevent the weight of the honey on the other side of the comb from breaking the comb. The combs should be turned to place the other side against the sides of the basket. Again, start the extractor slowly and gradually increase the speed to about 375 rpm for two or three minutes or until all of the honey has been thrown from that side of the combs. You can tell when the honey has been removed by watching it as it is sprayed against the side of the drum. When all of the honey has been extracted from the second side of the comb, again reverse the position of the combs and complete the extracting from the first side. The empty combs are then replaced in the super.

USE BOX OR PLATFORM

The extractor should be mounted on a box or platform so it is off the floor far enough so a pail may be placed under the spigot for receiving the liquid honey. It should preferably be anchored to the floor by three metal rods, or heavy wire, with a turnbuckle attached to each. Eye screws may be screwed to the floor for attaching the rods or wires. The extractor is likely to move about if it isn't anchored in this fashion. If it isn't feasible to anchor the extractor to the floor, it may be anchored to a 4 x 4 foot sheet of three-quarter-inch plywood which should be placed under the extractor.

The uncapping tub should also be set on a stand so the top is at a comfortable height for uncapping the combs. If you use a galvanized laundry tub, you may wish to make a hole in one corner to let the honey drain out into a pail. If you do, the opposite corner should be slightly elevated to allow the honey to drain to the other corner.

The holding tank should be on a platform which is at least high enough so a pail can be placed under the spigot. If you are going to put some of the honey into five-gallon cans, then the tank should be high enough to accommodate these. Plan for enough height so you can use a canning funnel on the top of the can. This funnel will help to steer the stream of honey into the can opening.

The supers of empty combs should be returned to the hives so the bees can remove the honey remaining in the combs, or add more honey if the honeyflow is still on. They should be returned toward evening to cut down on the chances of inciting robbing. They may be placed above the inner cover, if the hole in the inner cover is open, so the bees may come up through. After they are cleaned by the bees, they may be placed below the inner cover or put on other hives where they may be needed. This cleaning by the bees is especially important after the last of the honey crop has been extracted. Any honey remaining in the combs is likely to crystallize over the winter. These crystals will cause the next year's crop to crystallize in the combs before they are removed from the hive. It will be hard to extract the honey from these combs. The damage is permanent and will be repeated year after year unless the combs are melted.

STRAINING THE HONEY

The honey is drained out of the bottom of the extractor into a pail and is then strained through two thicknesses of cheesecloth, or one thickness of bolting cloth, into the holding tank. The cloth strainer should be moistened with water before use so the honey will pass through. It is well to place a wire household strainer under the spigot of the extractor to catch the larger bits of wax and any bees that may get into the honey. This will keep your cloth strainer a little cleaner and keep it more efficient. It is a good idea to have a wire screen mounted on top of the holding tank to support the cheesecloth strainer so it won't sag and allow coarser material to pass through it. A piece of window screen placed above the cloth strainer will catch the coarser material.

It is best to extract the honey as soon as it is removed from the hive while it is still warm. If this is not possible, then it should be stored in a warm room where the temperature may be maintained at between 80 and 90° F. It will extract easier, if kept warm, and less will be left adhering to the combs.

The honey should be left in the holding tank for two or three days to allow the air bubbles and bits of wax and propolis to rise to the top, where they may be skimmed off. It may then be drawn off into jars or into five-gallon (60-pound) cans for storage or further processing. Don't leave the honey in the tank too long or it may crystallize and become difficult to remove. Unheated honey will generally crystallize in a short time and this may spoil the appearance of the honey for some people. Crystallized honey is good honey, unless it has fermented, and some people prefer it in the crystallized state. However, crystallized honey is likely to ferment; which will spoil the flavor. Most people prefer the honey in the liquid state.

HEATING EXTRACTED HONEY

To maintain honey in the liquid state, and to prevent fermentation, it must be heated to 150° F. Before bottling, this may be accomplished by placing the honey in 60-pound cans and heating it in a water bath in a wash boiler. Most wash boilers will hold two cans. A wooden rack should be made to keep the cans about an inch off the bottom of the

---

HOW BEES MAKE HONEY AND WAX

The nectar that bees collect is generally half to three-fourths water. After nectar is carried into the hive, the bees evaporate most of the water from it. While evaporating the water, enzymes change the nectar into honey. Then the bees seal the honey into cells of the honeycomb.

Beeswax begins as a liquid made by glands on the underside of the worker bee's abdomen. As it is produced, it hardens into tiny wax scales. Worker bees then use this wax to build honeycomb.

Beekeepers often provide their bees with honeycomb foundation made of sheets of beeswax. This foundation fits into hive frames and becomes the base of the honeycomb. It enables bees to speed up comb construction, and it provides a pattern for building a straight and easy-to-remove honeycomb.

Beekeeping for Beginners (USDA)

boiler. Care must be taken to keep the honey from direct contact with the hot stove or it will become overheated and caramelize. That will detract from the flavor of the honey and make it darker colored. There should always be about an inch or more of water between the honey container and the source of heat. The water in the boiler should come up close to the top of the can to make for more uniform heating.

Use a candy or dairy thermometer for checking the temperature of the honey and stir it occasionally so the honey next to the outside of the container won't become overheated. Don't let the temperature go over 160° F. or there will be caramelization with deterioration in the quality of the honey. Gas, gasoline, or oil stoves are better for heating honey since the heat may be more easily adjusted and may be turned off when the desired temperature has been reached.

If honey has crystallized solidly in the can, it may take three or four hours to heat two 60-pound cans in a wash boiler to 150° F. The time required depends upon the starting temperature of the honey. Make certain that all of the crystals are melted or they will clog the strainer and also cause the honey to recrystallize in a short time. Stirring the honey every half-hour, after it has started to melt, will contribute to uniform heating and melting. Sometimes a mass of crystallized honey will sink to the bottom of the can where you won't notice it. Using a long stick for stirring will make you aware of such a mass. A flashlight will also help you to see it.

When the temperature of the honey has reached 150° F., it should be strained through 90-mesh strainer cloth. The cloth should be moistened with warm water before use and wrung as dry as possible. Otherwise, the honey will have difficulty in passing through. Wash a new cloth first to remove any sizing that it may contain. After the honey has been strained, it should be placed into suitable containers while still hot. Most honey is retailed in glass or plastic jars. Those holding one-half, one, two and five pounds are most popular. The twelve-ounce squeeze bottles are also good sellers. The jars of honey should be allowed to cool to room temperature before they are placed in cartons. The cardboard cartons hold in the heat and the result will be darker honey, if it isn't allowed to cool first.

CHAPTER 7

Fall and Winter Management

by Enoch H. Tompkins

Many beekeepers consider that their beekeeping year starts in the fall. The care that one gives his bees at this time of the year materially affects the size of the next year's honey crop. Without good fall management, there won't be as many bees available for gathering next year's honey. The following practices are designed to provide a maximum number of bees for the next year's honeyflow.

REQUEEN

Probably the most important thing that one can do in the fall is to make certain that each colony has a productive young queen. Such a queen will continue egg laying into the late fall and insure that there is a large number of young bees going into the winter cluster. Young bees will live longer than older bees and provide a good start for the following spring. A young queen will start laying eggs in larger quantities earlier in the spring and build up the labor force for the honeyflow. Older queens may be too nearly worn out from summer egg laying and not lay many eggs in the fall or the following spring. Some beekeepers requeen every year. At the least, queens that are two years old, or older, should be replaced. The best time for requeening is during a fall honeyflow. In the northern states, this will likely be during August or September when nectar is coming in from goldenrod and wild asters.

UNITE WEAK COLONIES

Only strong colonies should be wintered. They should have bees covering at least 20 deep combs at the time of the first killing frost. Any weak colonies should be united with strong colonies using a sheet of newspaper between and placing the weak colony on top. Poke a few pin holes through the paper so the bees can get hold of it and chew their way through.

CHECK WINTER STORES

Each colony should have a minimum of 60 pounds of honey for winter stores. They should be wintered in two deep hive bodies. The upper hive body (the food chamber) should be full of honey and pollen in dark brood combs. The lower hive body (the brood chamber) should be about half full of honey and pollen. There should be a minimum of the equivalent of one and one-half deep combs of pollen, preferably three to five well-filled combs in the northern states. Any colonies that don't have a minimum of 60 pounds of honey should be fed sugar syrup. Syrup for fall feeding should be made with two parts of sugar to one of water. Heat the water to the boiling point, remove it from the heat and stir in the sugar. Continue stirring until all the sugar is dissolved. The addition of one-quarter teaspoon of tartaric acid to each five pounds of sugar will prevent recrystallization of the syrup. Dissolve the acid in the water before adding the sugar. Feeding should be done so the bees may store the syrup before the weather gets too cold. They start to form their winter cluster when the temperature gets down to 57° F. When the temperature drops to 43° F. all of the bees will have joined the winter cluster. At this latter temperature, it is too cold for feeding the bees. See Chapter 5 on Spring Management of Overwintered Colonies for information on methods of feeding sugar syrup to bees.

CHECK FOR DISEASE

During the fall you should especially check for signs of disease. A colony that shows signs of American Foulbrood may not survive the winter. If it died or became weakened it would be a source of contamination to other colonies that would rob the honey and spread the disease to their hives. If you find evidence of disease, contact your state or county apiary inspector for advice on what to do.

If you do any feeding, it will be well to add *sulfathiazole* and *fumagillin* to the syrup for the prevention of American Foulbrood and the control of nosema. Dissolve the drugs in water before adding them to the syrup. Stir the syrup well to evenly distribute the drugs.

PROVIDE AN UPPER ENTRANCE

An upper entrance to the hive should be provided to remove moist air from the hive during the winter and for an emergency exit in case the lower entrance gets closed by snow, ice or dead bees.

Some inner covers have a notch cut in the rim for an upper entrance. This should be placed toward the front of the hive and the hole in the center of the inner cover should be left open. Make certain that the outer cover is pushed to the front so it doesn't block the exit.

Another method of providing an upper entrance is to drill a three-quarter inch hole in the front and upper part of the top super about half way between the handhold and the corner of the super. Don't drill it in, or under, the handhold or you may put your fingers on a bee sometime when you lift the super. If you do, you will likely get stung and may drop the super.

REMOVE QUEEN EXCLUDERS

All queen excluders should be removed from the hives before cold weather sets in. These should be placed in storage until they are needed next year. If a queen excluder is left between supers, it will prevent the queen from accompanying the cluster as it moves into the upper part of the hive. She will then die from the cold or from starvation and you will have a queenless hive when spring arrives.

Location of Apiary

PROVIDE A WINDBREAK

Your hives should be located where they have a windbreak to protect them from the prevailing winds. Such a windbreak can be a building, a row of trees and/or shrubs, a stone wall, a board fence or a snow fence.

A windbreak will cut down on the breezes that blow against the hive and dissipate the heat of the cluster. Bees have to take cleansing flights during the winter and a windbreak will help to make this possible by providing a warmer area for them to fly in.

Good air drainage is desirable. Your bees should be located in a place where the ground slopes away from the hive. This is desirable so that cold air and moisture will drain away from the hives rather than becoming stagnated in the area. The hives should be located where water will not collect and flood the hives or cause undue dampness.

A sunny location is best. It is better to have your hives located in a place where the sun will shine on them most of the time during the winter. The sun can warm up the hive and permit the bees to take cleansing flights or to move over onto fresh stores of honey. During prolonged periods of cold weather, the bees may starve, even though there is plenty of honey nearby, if they are unable to break their cluster and move over to the honey.

HOW TO MOVE A COLONY

If you need to move your bee colony, get the bees oriented to the new location. Unless you move at least several miles, the bees will find their way back to their old location.

If you want to move your bees only a few hundred yards, first take them several miles away and leave them for a week or more. After they get used to the distant location, move them to the nearby site you originally desired, and let them get oriented there.

Or, move the colony a few feet each day, until you have moved it to the location you want.

It is not advisable to move bees during the period of honey production. The honey already stored will add extra weight; new honeycomb may break loose; and you will disturb your bees and cause a slowdown in honey storage.

Night is the best time to move a colony. All the bees are inside then. If the weather is cold, you can completely close the hive entrance.

If the weather is unseasonably warm and the colony strong, do not seal the hive entrance. You might suffocate your bees, even if you seal them in for only an hour. Instead, cover the entrance and replace the cover of the hive with a fine screen.

Staple, crate, or tie the hive in advance so that parts cannot shift during the move.

Beekeeping for Beginners (USDA)

This snow fence provides an excellent wind break for these hives.

GUARD AGAINST MICE

When cold weather approaches, field mice start looking for a nice dry and preferably warm place to spend the winter. A beehive offers an ideal place. You should guard against their entry into the hives by using an entrance block. This reduces the entrance to a space about three-eighths by three inches, which is too small for mice to enter. If you plan to leave the full entrance open, then you should cover it with a strip of hardware cloth having three-eighths or one-half inch mesh. That will keep the mice out but will permit the bees to pass through.

This entrance block is positioned to provide minimum opening into hive.

Sheet of black building paper is wrapped around hive.

Enoch Tompkins cuts it, fits it into place.

Paper is tacked in place, entrance cleared.

Paper is doubled back away from entrance.

Paper at top is neatly folded into position.

And tightened so cover will fit in position.

Outer cover may not fit down over inner cover.

Rock holds it. See ventilation hole in front.

These two photos show a sure method of discouraging mice. In upper photo a strip of hardware cloth the same length as the hive entrance is bent into shape, then, in lower photo, is fitted into hive entrance. Cloth has ⅜-inch or ½-inch mesh so bees can enter easily.

If mice get into a hive, they will chew up the combs and annoy the bees. The bees leave the cluster to battle the mice and may not get back, if the weather is too cold. In the spring you may have a mouse nest and a dead colony of bees.

Packing for the Hive

In past years it was customary, in northern states, to provide winter insulation for the hives. Most beekeepers do not do this anymore. The same insulation that keeps the heat in also keeps the heat of the sun out. That prevents the sun from warming the hive so the bees may take a cleansing flight and/or move over onto fresh stores of honey on cold winter days. Bees do not try to heat up the entire hive during the winter. They are only concerned with keeping up the temperature within the cluster. Outside the cluster, the inside of the hive may be about as cold as the outside. Therefore, there is no need for insulating the hive.

Some northern beekeepers do wrap their hives in 15-pound black building paper. This absorbs the sun's rays on cold days and warms the hive enough so the bees may move onto fresh stores of honey which otherwise might not be possible. Some have even discarded this practice, feeling that the paper holds in too much moisture to the detriment of the bees. Many beekeepers don't pack or wrap at all, relying upon plenty of honey, lots of young bees, and a good windbreak to get their bees safely through the winter.

Winter Activities

Once you have your bees ready for winter, leave them alone until spring. Use the winter months to do some reading on beekeeping and to get your equipment ready for the next year. This is a good time to get your supers cleaned up and to put foundation in the frames for cut-comb honey. Order any extra equipment that you need and get it assembled and painted. You may wish to try your hand at making some of your equipment yourself. Complete instructions for making hives, supers and a honey extractor are available from Garden Way Publishing in Charlotte, VT 05445.

CHAPTER 8

Where Bees Can Be Kept

by *Garden Way Staff*

If we were to explore the world, looking at the places where bees are kept, we would realize that they are kept—and thrive—in what would seem to be the most hostile of regions.

Desert areas, lands near the Arctic Circle, urban areas, all of these and many more are places where bees are kept and honey is harvested.

For the gardener, one of the most pleasant places for a hive is in his garden. Here he can watch his bees flying industriously in and out, and he will be aware if something is wrong in the hive.

And a hive is more than a pleasant addition to the gardenscape. It's productive. What else can you "plant" that will add as much as 100 pounds of rich goodness in "crops" from a 20-square foot area?

The vegetable gardener shouldn't expect the pollinating bees to increase his yield noticeably. They may be attracted to his cucumbers, squash and a few other vegetables—and again they may fly over them, bee-lining for something more productive. One gardener recalls his pumpkin vines reached out to and finally surrounded one of his hives. A bumblebee or two visited the huge pumpkin blossoms daily, but his bees never paid any attention to them.

Should a gardener-beekeeper grow crops—a bed of clover or some herb favored by the bees—and thus get more honey? This is usually a futile experiment. Our Garden Wayer tried it with buckwheat covering a quarter of his garden. It was planted for two reasons, to provide buckwheat blossoms for the bees and to grow a crop that could be rototilled into the soil and thus enrich it.

He watched each morning as the buckwheat blossomed, and none of the bees from the nearby hive visited it. The experiment, however, was not a complete loss, since the soil was enriched by the green manure.

Where in the Garden?

If you put a hive or more in your garden, plan carefully where you place them, to avoid trouble later. Bees are irritated by the noises produced by the engines of rototillers and lawn mowers. They dislike traffic in front of their hives that disrupts their flights in and out.

Remembering these two points, the beekeeper-gardener will put some thought into the placement of the hives.

An ideal place is on the southern side of the garden, with the entrance pointed south and thus away from all of the traffic in the garden. This should be a spot reached by the sun in the early morning hours so the hive is warmed early—a thermostatic reveille for its inhabitants.

But what if that location is close to a road or sidewalk, or the favorite gathering place of the neighbors? It still can be used if handled properly. Bees, like planes at an airport, follow definite flight patterns, and usually take off directly from the hive, gradually gaining altitude. Their return is a reverse of this. That pattern can be changed, if something—a wall or tree or bush, is positioned to interfere and thus force them to gain altitude more quickly or to take a different route. Such planning will permit a hive to be placed in an ideal location and still not be a nuisance to others.

While a garden is a fitting spot for a hive, the suburban dweller without one may still keep bees. Proper planning will usually avoid future problems.

There are several rules for keeping hives in areas where home lots are small, rules that will avoid trouble with the neighbors.

1. Keep the hive in an inconspicuous place. Think of the worst possible place. It would be on the front lawn, near the sidewalk, and with

the hive entrance toward the sidewalk so the bees flew back and forth across the sidewalk and street. The first neighbor to complain would be the man with the large mongrel dog that befouls lawns, tips over trash cans and frightens children as the owner strongly defends the dog's civil rights. He will be joined by the parents whose child was stung last year by a wasp. Avoid this hassle. Place your hive in the rear of your home, and make it even more inconspicuous by hiding it among the shrubbery—but remember that need for early morning sun. Make certain the bees' flight pattern is not where it will interfere with pedestrian traffic.

2. Provide a water source for your bees. If you don't, the bees will find one, and it may be a neighbor's swimming pool or ornamental fountain or bird bath, and then those bees will be nuisances.

3. Be an ambassador for Beedom. The uninitiated are certain of but one fact: Bees sting. Teach them more about the complex society of the bees. Children particularly are fascinated once they have overcome their fears, and are good students when shown the inner workings of the hive. Also be generous with your honey, particularly if someone has been stung, whether your bees are involved or not.

Four-inch eaves trough filled with coarse gravel and equipped with float water-level control makes excellent water supply for bees. (USDA photograph)

Crops for the Bees

Most beekeepers agree it is not worth moving hives to an orchard, to reap the honey during the blossoming season. Professional beekeepers do move hives into orchards during that season, but the lure is the money paid the beekeeper for that pollinating service, rather than the honey harvested from such an orchard.

Apple orchardists are some of the largest commercial users of bees for pollination. Their relationships with beekeepers must be close, to protect each other's interests. The orchardist is interested in having the bees at the proper time, since pollination from one variety of apple to another is essential if there is to be a crop. And the beekeeper knows he is moving his hives early in the season and thus at a time when the hive is weak and should be left alone to gain strength. He must be certain his bees have adequate food at this critical period, and, above all, that they not be threatened with pesticides.

Such pollination efforts are best left to the professionals.

But this does not mean most bees will not pollinate many crops. The uninitiated think of bees happily going from rose to marigold to clover in their daily rounds. This is not the case. Bees prefer to work crop lands—acres of clover or alfalfa, for example—or trees, where there is a good supply of nectar or pollen, and where they are generally and unwittingly pollinating that crop.

Bees in the City

Bees are raised in cities. With their remarkable systems of seeking out food and communicating the presence of that food to the other gatherers in the hive, the bees can often produce a good honey crop in an area where man might see only sidewalks, streets and buildings.

A common location for hives in the city is on a rooftop. The beekeeper thus avoids the problems of the bees being close to pedestrian or vehicular traffic. A rooftop is not an ideal place for the bees, however, being far too hot in the summer and cold and open to the worst winds in the winter.

Planning is essential. The hive may thrive if it is placed so that it is well shaded from noon on during hot summer days. And if it is protected from wintery winds, the bees may prosper.

The city beekeeper should start modestly, with one or two hives, until he has determined the sources of food for his colonies. The proper trees in a nearby park, for example, may provide all that is needed. He should identify such sources before he expands his operation.

WHAT BEES NEED

Bees need four basic materials: Nectar, pollen, propolis, and water. They make honey out of nectar. They make pollen into beebread (food for young bees). They use propolis to seal cracks and waterproof their hive. They dilute honey with water before eating it, and they use water in their hive "air conditioning system."

Nectar

Bees can't make honey without nectar, a liquid sugary substance produced by flowers. It is the raw material of honey and the bees' main source of food.

Several hundred kinds of plants produce nectar, but only a few kinds are common enough, or produce enough nectar, to be considered major sources.

The best sources of nectar for producing surplus honey vary from place to place.

alfalfa	fireweed	sage
aster	goldenrod	sourwood
buckwheat	holly	star thistle
catclaw	horsemint	sumac
citrus fruit	locust	sweetclover
clover	mesquite	tuliptree
cotton	palmetto	tupelo
		willow

Pollen

As worker bees gather nectar from flowers, tiny particles of pollen stick to their bodies and are accumulated in pellets on their hind legs. This pollen is carried back to the hive where the bees store it as "beebread" in cells of the honeycomb.

An average-size colony of bees uses about 100 pounds of pollen each year. That is why you need to locate your colonies near good sources of pollen. Many wild flowers, ornamentals, weeds, shrubs, and trees will provide pollen. Some especially good sources are:

aster	dandelion	goldenrod	maple	poplar
corn	fruit blossoms	grasses	oak	willow

Beekeeping for Beginners (USDA)

CHAPTER 9

Development of
American Beehive

by Spencer M. Riedel, Jr., *apiculturist, Entomology
Research Division, Agricultural Research Service*

The modern American beehive, commonly called the *Langstroth hive,*
permits development of a strong colony and production of a large
honey crop. With it the beekeeper has control over the bees. It is
simple in design, mobile, light, durable, and economical. Its com-
ponents are interchangeable with those of other hives.

The dimensions for a Langstroth 10-frame hive are given in the illus-
tration. Basically the queen and her brood are confined in the brood
chamber. The worker bees can pass through the queen excluder and
store honey in the super. Since one shallow super is not enough space
for a populous colony, beekeepers often use several supers. The brood
chambers and supers are used interchangeably. If the combs contain
brood, the section is referred to as a brood chamber; if no brood is
present, it is called a super.

Reprinted in cooperation with Louisiana Agricultural Experiment Station from *Beekeeping in the United States,* USDA Agriculture Handbook 335.

Plans and dimensions of
Langstroth ten-frame beehive.

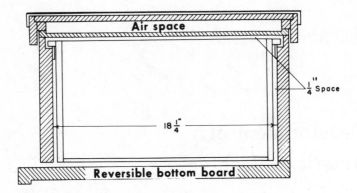

Air space

$\frac{1}{4}$" Space

$18\frac{1}{4}$"

Reversible bottom board

CROSS SECTION OF HIVE BODY AND FRAME

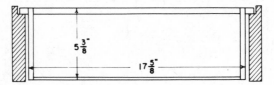

$5\frac{3}{8}$"

$17\frac{5}{8}$"

CROSS SECTION OF SHALLOW SUPER

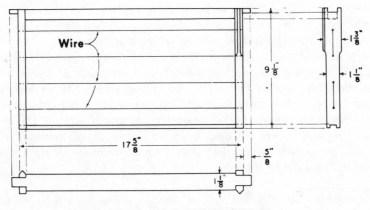

Wire

$9\frac{1}{8}$"

$1\frac{3}{8}$"

$1\frac{1}{8}$"

$17\frac{5}{8}$"

$\frac{5}{8}$"

$1\frac{1}{8}$"

SIDE, END, AND TOP ELEVATION OF FRAME

86

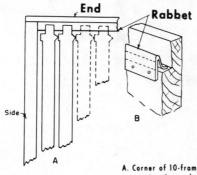

A. Corner of 10-frame hive body, showing construction and position of frames

B. Part of end of hive body, showing rabbet, which should be made of tin or galvanized iron

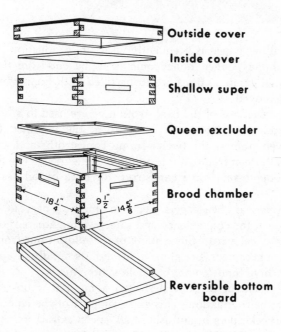

Outside cover

Inside cover

Shallow super

Queen excluder

Brood chamber

Reversible bottom board

(USDA drawing)

Straw skep, one of the earliest man-made hives.

The shallow super is frequently used in production of liquid or extracted honey, when honeycomb is to be cut from the frame, or when special comb sections are to be filled with honey and then individually removed.

Several variations of the Langstroth hive are used to a limited extent. They include the eight-frame Langstroth, the modified Dadant that holds eleven frames, the twelve-frame Langstroth, and a square hive that holds thirteen frames but is only 6½ inches deep. The most popular is the full-depth ten-frame Langstroth for both brood nest and honey storage.

Early-American beekeepers kept colonies of bees in hollow tree trunks, which for convenience and safety were gathered together in an apiary, or a "bee yard." Since many of these logs came from gum trees, the log hive became a "bee gum." This term is rapidly disappearing, but the terms "hive" and "colony" remain synonymous.

In 1921, Dadant & Sons perfected a method of inserting vertical wires into the foundation. This extra support of the comb was beneficial when beekeepers began using high-speed extractors. An aluminum comb was developed but was unacceptable to the bees. Both aluminum and plastic foundations have also been developed. They are embossed with the cells and coated on each side with beeswax. Since aluminum

Modern beehive cut away to show interior and placement of movable frames: Bottom, full-depth hive body; middle and top, shallow hive bodies. (USDA photograph)

is a good conductor of heat, it is not satisfactory in brood rearing areas of the hive. Neither of these permanent base materials is readily accepted by the bees unless properly installed in the frames and supplied to the bees during a heavy honeyflow.

The best foundation is still beeswax held firmly in the frame with embedded wires. The best type of hive under most conditions is the Langstroth ten-frame hive.

CHAPTER 10

Seasonal Colony Activity and Individual Bee Development

by Norbert M. Kauffeld, *apiculturist, Entomology Research Division, Agricultural Research Service*

Seasonal Colony Activity

Basically a colony of honey bees comprises a cluster of several thousand workers (sexually immature females), a queen (a sexually developed female), and, depending on the colony population, an indeterminate number of drones (sexually developed males). A colony normally has only one queen, whose sole function is egg laying and not colony government. The bees cluster loosely over several wax combs, the cells of which are used for the storage of honey (carbohydrate food) and pollen (protein food) and for the rearing of young bees to replace old adults.

The activities of a colony vary with the seasons. The period from September to December might be considered the beginning of a new

Reprinted in cooperation with Wisconsin Agricultural Experiment Station from *Beekeeping in the United States*, USDA Agriculture Handbook 335.

year for a colony of honey bees. The condition and activity of the colony at this time of year to a great degree affect its prosperity for the next year. In the fall the night temperatures decrease and the days shorten. Plant growth, nectar secretion, and flower characteristics respond in various ways that affect the activity of the bee colony.

Reduced incoming food causes reduced brood rearing and diminishing population. Depending on the age and egg-laying condition of the queen, the proportion of old bees in the colony will begin to decrease, since young bees emerging in the fall live longer. These young bees survive the winter while the old ones gradually die. Propolis collected from the buds of trees is used to seal all cracks and reduce the entrance to keep out cold air.

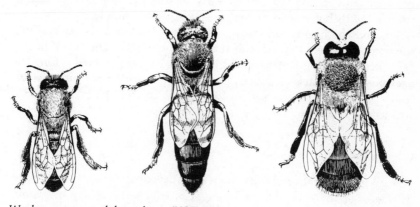

Worker, queen, and drone bees. (USDA photograph)

DEATH BY STARVATION

When nectar supplies in the field become scarce, the workers drag the drones out of the nest and do not let them return, causing death by starvation. This elimination of drones reduces the consumption of winter honey stores. If the queen is old, she may be killed by the workers during late summer and replaced by a young mated one before the drones disappear.

When the temperature drops into the 50's, the bees begin to form a tight cluster. Within this cluster the brood (consisting of eggs, larvae, and pupae) is kept warm, with heat generated by the activity of the bees feeding on the stored honey. The egg laying of the queen bee tapers off and may stop completely during October or November, even if pollen is stored in the combs. During the winter the colony is put to its severest test of endurance. Under subtropical, tropical, and mild winter conditions, egg laying and brood rearing never stop.

As temperatures drop, the bees draw closer together to conserve heat. The outer layer of bees is tightly compressed, insulating the bees within the cluster. As temperatures rise and fall, the cluster expands and contracts. The bees within the cluster have access to the food stores. During warm periods, the cluster shifts its position to cover new areas of comb containing food. An extremely prolonged cold spell can prohibit cluster movement, and the bees may starve to death only inches away from food.

The queen stays within the cluster and moves with it as it shifts position. About mid-January to early February if the colony is well supplied with honey and pollen, it will begin to stimulatively feed the queen and she begins egg laying. This new brood aids in replenishing the number of bees that have died during the winter. The extent of this early brood rearing is determined by pollen stores gathered during the previous fall. A lack of pollen delays brood rearing until more is collected from spring flowers, and the colony emerges from winter in a weakened condition.

POPULATION DECREASES

The colony population during the winter decreases because few young bees emerge to replace those dying. Colonies with plenty of young bees during the fall and an ample supply of pollen and honey for winter usually have a strong population in the spring.

During early spring the lengthening days and new sources of pollen and nectar stimulate brood rearing. The bees also gather water to liquefy thick or granulated honey in the preparation of brood food. Rarely will there be any drones within the colony at this time of the year.

Later in the spring the population of the colony expands rapidly and the proportion of young bees increases. As the population increases, the field-worker force also increases. It may collect nectar and pollen in greater amounts than is needed to maintain brood rearing, and surpluses of honey or pollen may accumulate. This, of course, will depend upon the availability of plants that produce pollen and nectar.

As the days lengthen and the temperature continues to increase, the cluster expands further and drones are produced. With increase in brood rearing and the accompanying increase in adult bees, the nest area of the colony becomes crowded. More bees will be evident at the entrance of the nest. A telltale sign of overcrowding is to see the bees crawl out and hang in a cluster around the entrance on a warm afternoon.

Combined with crowded conditions, the queen also increases drone egg laying in preparing for the natural division of the colony population by swarming. In addition to rearing workers and drones, the bees also prepare to rear a new queen. A few larvae that would normally develop into worker bees are fed a special food called royal jelly, their cells are altered to handle the larger queen, and her rate of development is speeded up. The number of queen cells produced will vary with races and strains of bees as well as individual colonies.

Regardless of its crowded condition, the colony will try to expand by building new combs if food and room are available. These new combs are generally used for the storage of honey, whereas the older combs are used for pollen storage and brood rearing.

SWARMING

When the first virgin queen is about ready to emerge and prior to the main nectar flow, the colony will swarm during the warmer hours of the day. The old queen and about half of the bees will rush en masse out the entrance hole. After flying around in the air for several minutes, they will cluster on a limb of a tree or similar object. This cluster usually remains an hour or so, depending on the time taken to find a new home by scouting bees. When it is found, the cluster breaks up and flies to it. On reaching the new location, combs are quickly constructed, brood rearing starts, and nectar and pollen are gathered. Swarming generally occurs in the Central States during May or June, although it can occur at almost any time from April to October.

SIGNS OF SWARMING

Here are some of the signs that your bees are planning to swarm.

• Queen cells are found in the hive.

• A large number of bees cluster on the outside of the hive. This also occurs in hot weather when there is little nectar to collect—and does not indicate a swarm.

• Little traffic in and out of the hive. If noticed, the beekeeper should inspect the hive interior. If the hive is abnormally crowded, it is an indication the field bees are not working, and is a certain sign they are planning to swarm.

The Garden Way Staff

After the swarm departs, the remaining bees in the parent colony continue their field work of collecting nectar, pollen, propolis, and water. They also care for the eggs, larvae, and food, guard the entrance, and build combs. Emerging drones are nurtured so that there will be a male population for mating the virgin queen. When she emerges from her cell, she eats honey, grooms herself for a short time, and then proceeds to look for rival queens within the colony. Mortal combat eliminates all queens except one. When the survivor is about a week old, she flies out to mate with one or more drones in the air. The drones die after mating, but the mated queen returns to the nest as the new queen mother. Nurse bees care for her, whereas prior to mating she was ignored. Within three or four days the mated queen begins egg laying.

During hot summer days the colony temperature must be held down to 93° F. The bees do this by gathering water and causing it to evaporate within the cluster by its exposure to air circulation.

DEFENSE AGAINST SWARMING

The best defense against swarming is to create ideal conditions for the bees so they will not wish to swarm.

Conditions that encourage bees to swarm include:

- Overcrowding, resulting in lack of space for the storage of honey or space for the broodnest. Add supers.

- High temperatures inside the hive. Shading helps.

- Insufficient ventilation. Improve it. This can be done by widening the entrance, or propping up one corner of the cover, or setting one or more supers off center to provide a crack of space between them.

The Garden Way Staff

During the early summer the colony reaches its peak population necessary for the collection of the greatest amount of nectar and the storage of honey for the coming winter. After reproduction, all colony activity is geared toward winter survival.

Summer is the most favorable time for the storage of food supplies. More food is available and warm weather permits free flight. The daylight period is the longest then, permitting maximum foraging, although rain or drought may reduce flight and the supply of nectar and pollen available in flowers. It is during the summer that stores are accumulated for winter. At times enough is stored that man can remove a portion and still leave ample for colony survival.

Individual Bee Development

The queen lays small, elongated, white eggs, one attached to the base of each cell. The egg remains in an upright position for three days, then a larva hatches.

The larva, a small white grub, is mass fed royal jelly by nurse bees for the first two days to such an extent that it literally floats in this food. For the next four days the worker larva is fed a less nutritious food at regular intervals (progressive feeding) by a different group of nurse bees. During these feeding periods it grows rapidly to several times its size at hatching, and soon occupies the greater part of its cell. Its cell is capped on the ninth day after the egg is laid. No feeding occurs after the cell is capped.

The next twelve days are spent in the capped cell in prepupal and pupal stages. Following the fifth or last larval molt, which occurs two or three days after the cell is capped, the distinct adult body parts appear, such as legs, antennae, wings, mouth parts, head, thorax, abdomen, and eyes. Initially the pupa is white and extremely soft textured, but as it grows, to emerge from the cell as an adult, it changes from white to a darker gray and finally to its adult color. Hair develops on the various body areas, and on the day before emergence it completes its sixth and final molt. On the twenty-first day the young worker bee cuts its way through the capping and crawls out of the cell as a physically adult honey bee.

The activity of a honey bee within the colony varies according to its age and the development of its internal glands. During the first three days following emergence, it aids in cleaning the nest area. From the fourth to seventh day after emergence, it feeds older larvae with mixtures of honey and pollen. About the seventh day after emergence its pharyngeal glands become developed and it produces royal jelly. As such a nurse bee, it mass feeds the young larvae or queen until it is ten to thirteen days old.

About the twelfth day of its life the wax glands, located between the fourth and seventh segments on the undersurface of the bee's abdomen, develop. Through some still unknown body process honey is eaten and converted into beeswax, which is secreted as small scales between these segments. The flakes of wax are kneaded by mandibular action into cell walls in the construction of combs. When the bee is about a week old, it takes short orientation flights during the afternoon to memorize its home location. Various activities such as the conversion of nectar into honey, packing pollen pellets in the cells, handling water brought in for air-conditioning, fanning for ventilation, and guarding the entrance are performed until about the twenty-first day, after which it begins foraging for food.

Adult honey bee workers live about six weeks during the peak activity periods of late spring, summer, and early fall. Most of them die in the field. Those that die in the nest are carried out some distance from the entrance and dropped to the ground.

Development of Drones

The development period of drones from egg stage to adult is twenty-four days compared with twenty-one days for workers and sixteen for queens. They are larger than workers, do not have stings, and are generally fed by the workers. Their activity consists merely in eating honey and making flights from the nest for possible mating with a virgin queen bee. They live about eight weeks during the active season but are killed off in the fall at any age.

The queen bee, which is longer than the worker or drone, has the shortest period of development from egg stage to adult. Her diet from the larval stage throughout her entire life seemingly is royal jelly, although she can eat honey herself. The nurse bees groom and care for her bodily needs from feeding to removal of feces. Her sole function within the colony is to lay eggs for the production of the necessary bee population needed to store sufficient food supplies for survival. Under natural conditions her life expectancy is two or three years, but some queens have been reported to live up to six years. She gives off a substance that has a stabilizing effect on the colony.

CHAPTER 11

Bee Behavior

by Stephen Taber III, *apiculturist, Entomology Research Division, Agricultural Research Service.*

Bee behavior refers to what bees do—as individuals and as a colony. By studying their behavior we may learn how to change it to our benefit.

Two very practical discoveries of bee behavior made our beekeeping of today possible. One was the previously mentioned discovery by Langstroth of bee space. The other was the discovery of G. M. Doolittle that large numbers of queens could be reared by transferring larvae to queen cells. The discovery of the "language" of bees and their use of polarized light for navigation have attracted considerable interest all over the world.

Much has been learned about the behavior of insects, including bees, in recent years. As an example, the term "pheromone" had not been coined in 1953, when Ribbands summarized the subject of bee behavior in his book, *The Behavior and Social Life of Honeybees.* A pheromone is a substance secreted by an animal that causes a specific reaction by another individual of the same species. Now many bee behavior activities can be explained as the effect of various pheromones. Unfortunately

---

Reprinted in cooperation with Arizona Agricultural Experiment Station from *Beekeeping in the United States,* USDA Agriculture Handbook 335.

for apiculture much of the study of pheromones has been connected with insects other than the honeybee.

Recently we have learned how certain bee behavior activities are inherited, and this information gives us a vast new tool to tailor-make the honeybee of our choice. Further studies should reveal other ways to change bees to produce specific strains for specific uses.

THE HONEYBEE COLONY

The physical makeup of a colony has been described. An additional requirement of a colony is a social pattern or organization, probably associated with a "social pheromone." It causes the bees to collect and store food for later use by other individuals. It causes them to maintain temperature control for community survival when individually all would perish. Individuals within the colony communicate with each other but not with bees of another colony. Certain bees in the colony will sting to repel an intruder even though the act causes their death. All of these, and perhaps many other organizational activities, are probably caused by the social pheromone.

There is no known governmental hierarchy giving orders for work to be done, but a definite effect on the colony is observed when the queen disappears. This effect seems to be associated with a complex material produced by the queen that we refer to as "queen substance." There is also evidence that the worker bees from ten to fifteen days old, who have largely completed their nursing and household duties but have not begun to forage, control the "governmental" structure. Just what controls them has not been determined.

These and many other factors make an organized colony out of the many thousands of individuals.

THE DOMICILE

When the swarm emerges from its old domicile and settles in a cluster on a tree, certain "scout bees" communicate to it the availability of other domiciles. At least some of these domiciles may have been located by the scout bees before the swarm emerged. The various scouts perform their dances on the cluster to indicate the direction, distance, and desirability of the domiciles. Eventually the cluster becomes united in its approval of a particular site. Then the swarm moves in a swirling mass of flying bees to it. Agreement is always unanimous.

WAX COMBS

As soon as the swarm enters the new domicile, food is collected and wax comb is built. The wax they use is secreted in tiny flakes from glands on the underside of the worker bee abdomen—but only when fresh food is available. The wax is molded by the bee mouth parts to form the intricate comb. The first comb consists of about twenty-five cells per square inch. This is the size worker bees are reared in. After there is a considerable amount of worker brood and increased population of bees, comb containing larger cells is built. This comb is used for rearing drones.

Although most references indicate that the presence of drone cells, even in small quantities, is objectionable, it is natural and normal for these cells to be present. They may have a morale-boosting or possibly some other beneficial effect.

The space between combs inside a colony varies greatly. Worker brood comb is about an inch thick. The open space between brood combs in a natural cluster is about three-eighths inch but varies from one-fourth to one inch.

The space between honey storage combs is much more uniform than between brood combs. The space left between capped honey cells is usually one-fourth inch or even less—room enough for one layer of bees.

Horizontal cells are used for storage of honey.

As the colony ages, the combs that were first used for rearing worker bees may be converted to honey storage comb; areas damaged in any way are rebuilt. These changes usually affect the bee space and result in combs being joined together with "brace" comb. Strains of bees show genetic variation in building these brace combs.

All these cells are horizontal or nearly so; vertical cells are used for rearing queens. Why horizontal cells are used for the rearing of brood and for honey and pollen storage, whereas vertical cells are built only for queen production is unknown.

FLIGHT BEHAVIOR

When several thousand bees and a queen are placed in new surroundings—which happens when the swarm enters its new domicile or a package of bees is installed, or a colony is moved to a new location—normal flight of some workers from the entrance may occur within minutes. If flowering plants are available, bees may be returning to the hive with pollen within an hour. Bees transferred by air from Hawaii to Louisiana and released at 11:30 a.m. were returning to the new location with pollen loads within an hour. Package bee buyers in the Northern States have noticed similar patterns in bees shipped from the South.

What causes this virtually instant foraging by bees? What determines whether they collect pollen, nectar, or water? If food and water in the hive are sufficient, why should they leave to forage? Does a pheromone or a hormone cause this flight activity? Answers to these questions can lead to directing bees to specific duties we desire accomplished.

CAUSES OF SWARMING

The basic causes of swarming are not understood. Usually a consistent sequence of events occurs prior to swarming, but the actual swarm may not emerge even if all the events occur.

Workers are first reared in great numbers; then comes a period marked by the rearing of both workers and drones. Large quantities of pollen and nectar are brought into the hive. This crowds the brood nest and restricts the number of eggs the queen is able to lay. From ten to fifty queen cells are produced, and shortly after the first one is sealed the swarm issues.

Swarms usually emerge from the hive in the late morning and often prior to the main flowering or honey-flow period. The more intense the flowering period, usually the more intense the swarming "fever" in an apiary. Some genetic lines of bees are much more prone to swarm than others.

One likely cause of swarming may be a change in production of certain pheromones by the queen bee that affects the workers. Swarming can often be discouraged by giving the bees more room above the brood nest area into which they can expand.

HOUSECLEANING

Certain waste material accumulates in a normal colony. Adult bees and immature forms may die. Wax scales, cappings from the cells of emerging bees, particles of pollen, and crystallized bits of honey drop to the floor of the hive. Intruders, such as wax moths, bees from other colonies, and predators, are killed and fall to the floor. Worker bees remove this debris from the hive. In some ants this behavior pattern is controlled by certain pheromone-like chemicals. When these chemicals are applied to live ants of the same colony, their sisters bodily haul them out to the graveyard. A pheromone of a similar nature may occur in the honey bee colony.

The cleaning behavior associated with removal of larvae and pupae that have died of American Foulbrood in the cells is known to be genetically controlled by two genes. It is modified by honeyflow conditions. What makes this discovery important is not only that it should help in developing bees more resistant to this and related bee diseases, but also that we can now expect to find that many of the other behavior characteristics of bees can be modified to suit special needs, with mutual benefit to beekeepers and farmers.

BROOD REARING

The seasonal changes in egg production by the queen and the subsequent rearing of brood by the workers are thought to be mainly dependent on the supply and abundance of nitrogenous and carbohydrate food. However, brood rearing slows down and stops in the fall, then starts again during midwinter in both Louisiana and Wisconsin, states geographically and climatically dissimilar. The buildup of fat bodies in fall and winter bees coincides with similar physiological changes in other species of insects, such as the boll weevil going into a period of diapause or inactivity. Initiation or control of diapause has been shown to be regulated by quantity and quality of light.

An important aspect of bee behavior is the possible effect of the twenty-four-hour or circadian rhythm on daily brood rearing activities.

Much of the brood rearing activity cycle of colonies during the year can be related to the available forage or plants in bloom. Colonies in different parts of the country show fluctuations in brood production and in their subsequent colony populations that are coordinated with forage conditions.

Colonies in the Northeastern and Midwestern United States increase their brood rearing and bee populations during maple-dandelion-fruit bloom periods and show an intense swarming tendency prior to bloom of clovers. The brood and adult populations generally and slowly decline during late summer, but increase in areas with abundant acreages of aster, goldenrod, and smartweed. A severe decline in brood rearing follows the first killing frost. But in areas of New York, Pennsylvania, and Wisconsin, where the late flowering buckwheat is grown, greatly increased brood rearing, population increase, and even swarming occur in the early fall.

Beekeepers in the Western United States depend on flowering of irrigated agricultural crops and indigenous plants stimulated by infrequent rains for buildup of brood and adult bee populations. In the Southwest the farmers on irrigated land need bees on specific, predictable dates for pollination of their crops. The beekeeper who must supply these bees is dependent on unpredictable rainfall to provide natural food for his colonies to develop brood and bee populations needed for this service.

In the Southeastern States many different floral sources bloom abundantly at different but specific times of the year. Brood rearing and colony expansion commence early, and high populations are maintained with relatively little difficulty from swarming. Brood rearing usually declines somewhat during high summer temperatures when few flowering plants are available for bee activity, then picks up again late in the summer and continues until a killing frost.

Variation in the quantity of brood reared by colonies of bees nationwide is obviously affected by the immediate local environment. Yet with these apparent differences there are also similarities associated with photoperiods or other external stimuli rather than the quantity of food available.

TEMPER AND ITS CONTROL

The temper or gentleness of bees determines their inclination to sting. Many factors affect their temper, including the genetics or inheritance of the bee, environment of the hive, and manipulation of the colony by the beekeeper. The temper of the colony can be temporarily controlled by man with a certain amount of smoke. Exactly why and how smoke affects bees is unknown, even though it has been used by beekeepers worldwide for over a hundred years. The amount of smoke needed to control the temper of a colony varies with time, temperature, and various other external factors, as well as with the inherent gentleness of the colony. Ruthless manipulations that injure or kill bees create more need for smoke than careful manipulations. The right amount of smoke to use on a colony is learned only by experience.

TEMPERATURE CONTROL BY BEES

Bees are paradoxical in that they are individually cold-blooded insects but collectively behave like a warm-blooded animal. During periods of high temperature they bring in water, which on evaporation cools the cluster to the desired temperature. During periods of low temperature they cluster tightly and generate and conserve heat to hold the temperature up so that the center of the bee cluster rarely gets colder than 80° F. Neither the exact mechanism of cooling nor the building up of heat in the cluster is understood. None of the proposed theories has been proved.

COLONY MORALE

"Colony morale" generally refers to the well-being of the colony. If the colony morale is good, the bees are doing what is desired of them, including increasing the colony population, making honey, and pollinating flowers. Many factors seem to affect colony morale. For example, if the queen is removed from a colony during a honeyflow, the daily weight gains immediately decrease, although the bee population for the next three weeks is unaltered. Also when a colony is preparing to swarm, the bees practically stop gathering pollen and nectar. Improper manipulations or external environment also affects colony morale.

KNOWN PHEROMONE ACTIVITY

Some of the bee colony pheromones and their biological action are known to beekeepers. The Nasanov or scent gland was described over a hundred years ago. The biological behavior activity of its pheromone is best seen when a swarm is hived. When the bees first enter the new domicile, some bees stand near the entrance and fan. At the same time, they turn the abdominal tip downward to expose a small, wet, white material on top of the end of the abdomen. This seems to affect the other bees, for within several minutes all will have settled and entered the new hive. When bees find a new source of food, they mark it with the same scent gland. A Canadian research team has recently reported isolation and identification of this pheromone.

Colony odor refers to the odor of an individual colony. Because each colony odor is different, the colonies cannot be combined in the same hive without fighting and killing one another. Colony odor probably results from a combination of endogenous (pheromone or pheromone-like) and exogenous (food accumulation and food interchange) materials in each hive and seems to be recognizably different for every colony. The usual procedure for the beekeeper when colonies are to be combined is to place a newspaper between the two sets of bees. By the time the bees have eaten through and disposed of the newspaper, their odors have intermingled and become indistinguishable. During heavy honey-flows, differences between colonies seem to disappear, and colonies can be united without difficulty.

One of the most interesting and complex pheromones, originally termed "queen substance," is now believed to be a complex of different chemical pheromone compounds, which stimulates a large number of complex behavior responses. Its presence in virgin queens in flight attracts the drone for mating from an unknown distance. Its presence in virgin and mated queens prevents the ovaries of the worker bee within the hive from developing and the worker bees from building queen cells. It keeps swarming bees near the queen. Its decrease is a cause of swarm preparation or supersedure. Queen substance is produced in glands in the queen's head. If these glands are removed, the queen presumably produces no more queen substance, yet the queen and the colony go on much as they did before. The quantity and quality of queen substance vary in virgins and mated queens of different ages.

The alarm or sting pheromone, which also may be a complex of pheromones, has been tentatively identified by a Canadian research team. When a bee stings, other bees in the immediate vicinity also try to sting the same spot. In this case the sting pheromone provokes other bees to sting the same place, and they in turn provoke additional stings ad infinitum. Smoke blown onto the area seems to neutralize this effect.

Whether the genetic basis for a difference in temper is in the quantity of the alarm pheromone or pheromones produced or in the reception organs of other bees is unknown.

More on Bee Communication

Besides the known and possible methods of bee communication or language that have been mentioned involving the chemical pheromones, there are, of course, others. The best known is the so-called "dance" of the returned forager bee. This dance tells other bees precisely in which direction to search for food and how far they should fly, the type of food they will find, and the relative quantity. Minor genetical variations in individual dance behavior exist. How this genetic variance can be used in an economic way is unknown. The cluster itself and individual bees in the field make subtle noises or sounds, all of which are not at present understood. One such sound seems to tell the bees to stay home. Utilization of this sound could lead to protection of bees from harmful pesticides.

Experienced beekeepers recognize a difference in sound between a colony with a queen and one without, between a "mad" bee and an undisturbed one, and between normal and cold or starving bees. Individual queens and even worker bees emit squeaky sounds called "piping" and "quacking."

When a bee returns from a foraging trip and performs a dance, she communicates the kind of "plant" or "flower" on which she was foraging by releasing a taste or the perfume of the flower through nectar regurgitation or on body hairs. This has promoted other experiments designed to train or force bees to collect or work a desired crop. These experiments have so far been unsuccessful. The reason for the failures may well be that the bee language code has not been completely translated. We are still unable to "talk" effectively to the bees and "tell" them what we want done.

Bees also recognize and are guided by different colors but are unable to communicate these colors. Their eyes are receptive to the polarization of the light in the sky and this aids in their navigation.

AGE LEVELS AND WORK HABITS

The bee is adaptable to many environments. Honey bees native only to Europe, Asia, and Africa have adapted well to all but the polar regions of the world. Part of this adaptability lies in the capacity of the individual bee to "sense" what must be done, probably through reception of colony pheromones, then to perform the necessary duty.

Under normal conditions there are bees of all ages in the hive. The age of the bee determines in general its daily activity. However, when conditions become abnormal, age is no longer a criterion of duties. If a colony is made up of old bees, for instance, some of them will become physiologically young, and conversely if it is made up of all young bees, some will become physiologically old and take on duties normal to such bees. The reasons and causes for bees changing work in this manner are unknown.

Many beekeepers set colony on scales to measure inflow of honey.

BEHAVIOR OUTSIDE THE HIVE

Genetically we have found that some bees produce more honey than others, but we do not know why. The individual bee may collect more because of its own genetic inheritance. The colony may store more honey because of the queen's inherited ability to lay more eggs, resulting in a greater total population of bees in the hive, or because the bees are inherently longer lived. These activities would have the same gross effect of increasing the population of the colony.

We can, as "bee controllers," affect the bee's environment in conjunction with its inheritance. There is no reason to believe that bees work better when in want. Since evidence indicates the reverse, there should always be an ample supply of reserve honey in the hives.

Another environmental factor is colony manipulation. A beekeeper's disturbance of the colony during the honeyflow results in a marked decrease in the amount of honey stored for that day and even the following day. Colonies of bees should not be needlessly disturbed; however, manipulation to give them extra room to expand the cluster or store food is necessary.

POLLINATION—CROP PROBLEMS

A problem associated with bee behavior is in getting bees to work on a particular field. Growers of seed and fruit crops rent bees in ever-increasing quantities. They want the bees they rent to stay on their acreage.

One method being explored at the present time takes advantage of inherited behavior differences. Today a bee is being bred that is specifically suitable for alfalfa pollination. Bees specifically designed genetically to pollinate certain other crops will eventually be developed.

In conjunction with these is genetic variability in the attractiveness of plants to bees. Nectar and pollen availability in plants can be accidentally eliminated by breeding. When this occurs, there is a loss of a potential honey crop, but more important can be the loss of a seed or fruit crop because the plant no longer attracts pollinators. A beekeeper visiting Europe and seeing the large acreages of potatoes cannot help but wonder what would happen to the European bee industry if a strain or variety of potatoes were developed that would yield pollen and nectar for bees.

CONTRADICTORY BEHAVIOR

Probably the biggest problem in adequate pollination that must be solved pertains to the foraging activity of the bees. Individual bees usually confine their foraging area to a relatively small block of a few square yards or to a single tree. On the other hand, the foraging area of a colony comprises entire square miles—a circle with a radius approximating two miles—with the collective effect that foraging bees shift about, readily affected by the dance intensity of other returning foragers.

This contradiction could be resolved by breeding bees with greatly expanded individual foraging areas but with a reduced radius of colony activity, possibly three-fourths mile.

CONTROL OF FORAGING

A major goal is to control the foraging bee and get it to more effectively pollinate the plant. Bees should be attracted to areas where they are desired, e.g., pollination areas, and repelled from areas where there is danger to them from insecticides or where they endanger people. Work with other insects—both social and nonsocial—indicates that this could be accomplished by chemical and physical means.

Actually beekeepers have been using various chemicals to control bees in the hive for some time. Smoke, a combination of chemicals in gaseous form, and various chemical repellent compounds have been used for years to repel bees.

There is considerable evidence that different plant species produce varying attractant compounds associated with their nectar and pollen. Bees are highly attracted to the scent of recently extracted honeycomb and to the scent of honey being extracted or heated. During periods of intense honeyflow, bees much prefer collecting nectar of lower sugar concentration than exposed honey. Obviously certain chemical scents associated with certain flowers and to some extent incorporated in the collected honey are highly attractive to bees.

Certain chemical extracts of pollen strongly indicate that some pollens contain compounds that stimulate collection response in bees. Isolation and identification of these bee-attractive compounds and the application of the attractant to areas requiring high bee populations for pollination should attract bees to the area.

Synthetic chemical compounds, not necessarily attractive, can probably be used if the bees are properly trained. Bees associate certain smells with food sources and communicate the smell at the food source to the colony. Isolated chemicals, which may or may not have anything to do with pheromones produced by the bee or attractant compounds produced by plants, profoundly affect bee behavior. Some of these chemicals cause the bees to fight or flee from the source, but others attract the bees.

Of immediate concern is the need for a repellent that can be applied to a field to drive off all pollinators while a pesticide is applied. Then, it is hoped, when the toxic effect of the spray has disappeared, the repellent will dissipate and the bees return to work.

Research should not be confined to chemicals alone, but should be shared equally with various physical factors that can possibly attract or repel bees. In other entomological fields the research on physical methods of controlling insects is receiving intensive investigation. Different insects respond in differing ways, they are attracted to certain light wavelengths and repelled by others. Night-flying moths are repelled or go into defensive maneuvers because of bat sonar signals, whereas crickets and other members of their insect group can be collected by reproducing certain stridulations.

BEHAVIOR ACTIVITIES OF BEES

The drone. The time of day that drones fly in search of a mate depends on many factors, such as the geographic location, day length, and temperature. Drones usually fly from the hive in large numbers between 11:00 a.m. and 4:30 p.m. Morning or early afternoon flights last about two or three hours. Later flights are shorter. An individual drone flies for about fifteen minutes before returning to the hive and may fly four or five times in one day.

When flying, drones seem to congregate in "mating areas," which may serve to attract virgin queens. These areas are usually less than 100 feet from the ground and seem to be associated with land terrain.

Some control of the time of drone flight can be obtained. The usual afternoon flight can be prevented by placing the drone-containing colony in a cool, dark room. On the following day the drones will fly several hours earlier than normal.

The queen. The virgin queen becomes sexually mature about five days after emergence. She is relatively quiet in the morning and most active in the afternoon. She may begin her mating flights five or six days after emergence and may go on a number of them over several days. Mating with eight to twelve drones will stock her spermatheca with six or seven million sperm. She will begin to lay in two to five days and may continue for a year or more.

A young, fully mated queen rarely lays drone eggs before she is several months old. After that time she seems capable of controlling the sex of the eggs by laying either fertilized or nonfertilized eggs.

The worker bees occasionally kill their queen. More frequently they will kill a newly introduced or virgin queen. To do this, fifteen or twenty worker bees collect about her in a tight ball until she starves. It has generally been thought that bees "balled" strange or introduced queens because they did not have the proper "colony odor." The reason for balling is probably more complicated than that, because bees will occasionally ball their own queen. Even if the ball is broken up, the queen seldom survives. If it is broken up or after the queen is dead, the stimulus is powerful enough that the bees taking part in the queen balling are sometimes subsequently balled themselves by other bees.

LONG LIVE THE QUEEN

The beginning beekeeper will marvel at the decisions made in the complex, efficient operation of the hive. One of these is in making certain the hive has a queen—and a productive one.

When the queen is missing or is judged inadequate or defective by the colony, she is replaced.

First step is for the bees to build one or more queen-cells, usually along the edge of a comb. These are far larger than the usual cells, and are readily recognized by their peanut-shell shapes, hanging downward. The cell is either built around an egg or larva, or a worker egg is deposited in the cell after it is built.

The queen, in her fifteen days of maturing from egg to insect, is fed royal jelly by the worker bees during her entire period of growth, and it is this rich, abundant diet that produces a queen rather than a worker bee.

The Garden Way Staff

CHAPTER 12

Honey Bee Nutrition

by L. N. Standifer, *apiculturist, Entomology Research Division, Agricultural Research Service*

A general knowledge of honeybee nutrition aids in understanding how the individual bee grows and how the colony develops and maintains itself.

The anatomical and vital physiological systems usually associated with living animals are present in the honeybee. Food enters the alimentary canal of the adult bee by way of the mouth and long tube-like esophagus, which extends through the thorax and into the abdomen, where it enlarges to form the honey stomach. Nectar is transported in the honey stomach from the flower to the hive. Immediately behind the honey stomach is the proventricular valve. It retains the nectar load in the honey stomach, lets food pass into the midgut, but prevents food from returning. The midgut or ventriculus is a relatively large segment of the alimentary canal, lined on the inside with the peritrophic membrane. Beyond the midgut are the short, small intestine, the large intestine or rectum, and finally the anus.

The adult honey bee has six sets of paired glands located in the head

Reprinted in cooperation with Arizona Agricultural Experiment Station from *Beekeeping in the United States*, USDA Agriculture Handbook 335.

and thorax. The labial glands are generally believed to be associated with the alimentary canal. They deliver their secretions at the base of the labrum. Their function is dependent on the age of the bee and on the work in which it is engaged.

The hypopharyngeal or brood food glands produce the food called royal jelly. Most of the larval food comes from these glands. They also supply the food for the adult queen and possibly adult drones. They are fully developed only in the worker bee. Royal jelly is milky in appearance, slightly acid, and rich in digestive enzymes, proteins, carbohydrates, fats, and vitamins.

The mandibular glands are saclike, single structures located immediately above the mandibles. They are extremely large in the queen, smaller in the worker, and vestigial in the drone. They do not vary in size with age or occupation. However, if the newly emerged bee does not consume adequate protein during the first few days of adult life, these glands do not develop fully.

The secretions from these glands in queens contain a compound called queen substance, which is essential for social unity of the colony. This substance is not in the glandular secretions of worker bees. Workers probably use the secretion from their glands to prepare and manipulate wax for building comb. It may also be used to soften the pupal cocoons of bees and is found in royal jelly.

The primary function of the postcerebral glands, located in the back of the head, is to provide the enzymes necessary for the digestion of foodstuff consumed by the bee.

The thoracic or salivary glands in the anterior part of the thorax secrete a carbohydrate-splitting enzyme, invertase, in large amounts.

The function of the postgenal glands in the lower inner wall of the head and the sublingual glands at the base of the bee tongue is unknown.

Two pairs of rectal glands or pellets on the sides of the rectum are associated with fat absorption.

Beeswax is secreted by specialized cells called wax glands on the underside of the abdomen of the worker bee. Generally wax glands become fully developed about the fifteenth day of adult life. Secreted beeswax appears as thin, delicate scales or flakes. It is a byproduct of metabolism and directly follows the digestion of large quantities of nectar or other sugars. The bees use wax to form the cells of the comb, in which food is stored and brood is reared.

When wax is needed for comb building, the sixteen- to twenty-four-day-old bees fill their honey stomachs, then hang together in vertical sheets or festoons. The wax scales are secreted, then removed with the hindlegs, and passed forward to the mouth, where they are worked by the mandibles and subsequently applied to the edge of the comb. The wax glands shrink and become nonfunctional between the twentieth and twenty-fifth day of adult life, or about the time worker bees become field-foraging bees.

Digestion

The movement of pollen through the alimentary canal of the adult bee reveals something of the digestive process. Ten minutes after the material is fed to the bee, the pollen grains are clustered at the proventriculus. Thirty minutes after feeding, they are within the peritrophic membrane in the forepart of the ventriculus. Ninety minutes after feeding, the peritrophic membrane-enwrapped pollen mass enters the anterior or small intestine. At the end of two hours the pollen is within the small intestine or just entering the large intestine or rectum. The peri-

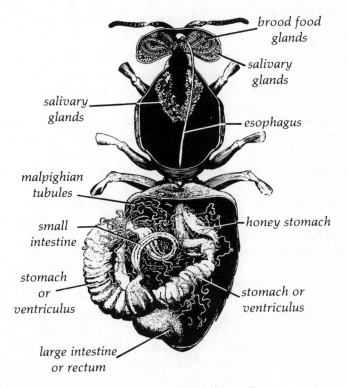

Alimentary canal of adult worker bee. (USDA drawing)

trophic membrane that encircles the pollen grains in the midgut persists in the rectum for a considerable time before all is voided in the feces.

The fatty acids in pollen are made water soluble by neutralization with alkalies in the alimentary canal secretion. The proteins are broken down into peptides, and these are further hydrolyzed into amino acids.

The lipids of bee food occur chiefly in pollen. The lipid-splitting enzyme lipase is abundant in the midgut of the adult worker and drone. Its value in digesting the lipids is unknown. In higher animals lipids are digested by lipase or esterase into free fatty acids and glycerol.

Nutritional Requirements

Honey bees and other insects have no unusual nutritional requirements. They require carbohydrates, proteins, fats, minerals, vitamins, and water for growth, development, maintenance, and reproduction. Nectar and honeydew are the chief sources of supply for carbohydrates in the diet of bees, and pollen furnishes all the other indispensable constituents.

Adult bees can live on the carbohydrates glucose, fructose, sucrose, trehalose, maltose, and melezitose. They cannot utilize the carbohydrates galactose, mannose, lactose, raffinose, dextrin, insulin, rhamnose, xylose, or arabinose. Beekeepers often feed sucrose if a shortage of nectar or honey exists. Bees also utilize fruit juices and certain occasional plant juices. A small amount of carbohydrates is also obtained from pollen. The adult bee can survive on carbohydrates; however, proteins, fats, minerals, and vitamins are necessary in rearing the immature stages.

Proteins of a precise quality and definite amino acid composition are required for optimum growth and development of the brood food-producing hypopharyngeal glands and no doubt others. When nursing duties are finished (between tenth and fourteenth day of adult life) and field duties are undertaken, the requirements for protein decrease, and the chief dietary constituent becomes the carbohydrates obtained from nectars and honey.

Fats, like carbohydrates, are also used as sources of energy. Bees probably require and utilize some of the fats in pollen they consume. However, chemical analyses of feces show that large amounts of fats in the pollen consumed by bees pass through the digestive tract and are not utilized.

Water is necessary in the diet for diluting concentrated honey. It is also used in air-conditioning the cluster. Normally bees do not store water as they do nectar and pollen. It is collected only when needed.

The nutritional value of the enzymes, coenzymes, and pigments found in pollen is largely unknown.

Sources and Composition of Foods

NECTAR

When nectar is collected, it may contain from 5 to 75 percent soluble solids (sugars), most of which is in the 25 to 40 percent range; the remainder is water. The primary sugars are sucrose, glucose, and fructose. As the nectar is manipulated and finally stored as honey, much of the sucrose is inverted to glucose and fructose, usually in about equal amounts.

HONEYDEW

Various species of insects secrete a material called honeydew, which bees collect and store in the comb. Honeydew has a high percentage of dextrins and melezitose. It is generally considered a poor source of carbohydrates for bees.

A frame well filled with honey.

PLANT JUICES

Bees often collect juices from overripe fruit and various plant exudates that are rich in sucrose or related sugars. They usually do this only when nectar is not available.

POLLEN

Pollen furnishes all the other indispensable constituents of the diet, except water, that are required for vital activity, including rearing young bees. Not all pollens are alike nutritionally, and bees grow and develop better on some than on others.

In nature bees generally utilize a mixture of pollens in their diet. This is eaten by adult bees and is fed to worker and drone larvae after they are three days old. Consumption and digestion of pollen by adult bees are essential, since they can only produce brood food from pollen that they have eaten. This brood food, or royal jelly, is fed to all larvae the first three days of life and to the queen bee throughout her larval and adult life. Royal jelly has the following approximate chemical composition (percent): Water 66, dry matter 34; of the latter, carbohydrate 13, protein 12, fat 5, ash 1, and undetermined matter including vitamins, enzymes, and coenzymes 3.

Bee-collected pollens are comparatively rich in the carbohydrates. Reducing sugars range from 15 to 43 percent, with an average of about 29 percent. The glucose, fructose, sucrose, raffinose, and stachyose content is not significant, although the bees apparently utilize those that are available. Corn pollen, for example, has a high starch content. The pollen shell is not utilized by bees, but is eliminated with the feces after the internal matter has been removed by digestive processes.

The protein value of pollen varies from 10 to 36 percent. The amino acid content of average pollen and sweet corn pollen with a crude protein of 26.3 and 26.9 percent, respectively, is shown in Table 1.

All the amino acids in Table 1, except threonine, are essential for normal growth of the young adult bee. With the exception of histidine and perhaps arginine, they cannot be synthesized by bees and must be obtained from the pollens consumed.

Bees also occasionally collect spores and store them as pollen. Although spores can be utilized as a proteinaceous food, they do not stimulate brood rearing and are generally considered a poor substitute for pollen.

TABLE 1. AMINO ACID CONTENT OF AVERAGE
POLLEN AND SWEET CORN POLLEN, EXPRESSED
AS PERCENT OF CRUDE PROTEIN

COMPONENT	*AVERAGE POLLEN*	*SWEET CORN POLLEN*
	Percent	*Percent*
Arginine	5.3	4.7
Histidine	2.5	1.5
Isoleucine	5.1	4.7
Leucine	7.1	5.6
Lysine	6.4	5.7
Methionine	1.9	1.7
Phenylalanine	4.1	3.5
Threonine	4.1	4.6
Tryptophane	1.4	1.6
Valine	5.8	6.0

Other constituents of pollen are as follows:

Constituent	*Amount*	*Constituent*	*Amount*
	Percent		*Micrograms per gram identified*
Fats	1.3–19.7	Vitamins:	
Minerals (ash):		Ascorbic acid	131.0–721.0
Calcium	1.0–15.0	Biotin	0.19–.73
Chlorine	0.6–0.9	D	0.2–0.6
Copper	0.05–.08	E	0–0.32
Iron	0.01–12.0	Folic acid	3.4–6.8
Magnesium	1.0–12.0	Inositol	0.3–31.3
Phosphorus	0.6–21.6	Nicotinic acid	37.4–107.7
Potassium	20.0–45.0	Pantothenic acid	3.8–28.7
Silicon	2.0–10.4	Pyridoxine	2.8–9.7
Sulfur	0.8–1.6	Riboflavin	4.7–17.1
		Thiamine	1.1–11.6

Artificial Diets

The beekeeper can supplement the diet of nectar or honey with sucrose.
This is usually mixed with about an equal amount of water and fed as
sirup.

There is no substitute for pollen. Various materials, including brewer's yeast, soybean flour, dry skim milk, and egg albumin, mixed with honey or sugar water, have been fed to bees, but the colony stimulation is minor compared to that derived from fresh pollen. The addition of dried pollen trapped from colonies earlier increases the stimulation slightly.

The two most commonly used artificial diets are the pollen supplement diet and the pollen substitute diet. Their composition is as follows:

Materials	Percent
Pollen supplement:	
Sugar-water (2 parts sugar to 1 part water by weight)	67
Pollen-soy mix (1 part fresh dry pollen to 3 parts soybean flour by weight)	33
Pollen substitute (dry mix):	
Soybean flour	20
Casein	30
Brewer's yeast	20
Dry skim milk	20
Dried egg yolk	10

Few problems facing the apiculture industry today require immediate research attention as much as the development of an artificial or chemically defined diet for honey bees as a substitute food for pollen. Work on nutrition and physiology of the honey bee may soon lead to an artificial diet for bees.

CHAPTER 13

Managing Colonies for High Honey Yields

by F. E. Moeller, *apiculturist, Entomology Research Division, Agricultural Research Service*

Colonies of bees existing in the wild away from man's control will produce small surplus crops of honey above their requirements for survival. Such surplus will vary depending on the region or locality, but will seldom exceed 30 to 35 pounds. In the same area subject to the same nectar resources, colonies properly managed will produce surplus honey crops of 150 to 200 pounds. Intensive two-queen colony management can often result in surplus crops of 500 pounds or more with the same resources available. The key to these differences is management.

Proper management employs practices that harmonize with the normal behavior of bees and brings the colony to its maximum population strength at the start of the bloom of major nectar-producing plants. Management practices are similar in basic principle wherever bees are kept and vary only as regards timing for the desired nectar source of the region or locality concerned.

Reprinted in cooperation with Wisconsin Agricultural Experiment Station from *Beekeeping in the United States*, USDA Agriculture Handbook 335.

Regardless of the type of hives or equipment used, proper management aims at providing colonies with unrestricted room for brood rearing, ripening of nectar, and storage of honey, plus provision of adequate food requirements, both pollen and honey, for the time of year concerned. Swarming is minimized and the storing instinct encouraged when proper management is used.

Preparing Colony for New Season

In the temperate regions of the northern hemisphere, August to October is the time when the beekeeper prepares his colonies for the coming year. This is when the major honeyflows are usually past and the bees must be made ready for the coming winter.

All queens of questionable performance with only a small amount of brood or irregular pattern should be replaced. Frequently the bees of the colony will replace or supersede queens of subnormal performance even before the beekeeper senses a problem. Some queens may be satisfactory in their second year; queens less than a year old are usually best.

To requeen a colony, certain principles of queen acceptance must be borne in mind: (1) Strong colonies more reluctantly accept a queen than weaker ones, (2) temperamental bees are more reluctant to accept a new queen than gentle bees, (3) young bees accept a queen more readily than older bees, (4) the colony to be requeened should first be made queenless, and (5) the queen to be introduced should be in egg-laying condition.

There is less risk in requeening a colony by giving it a laying queen with some of her own brood and bees than by giving it a queen in a shipping cage. A new or valuable queen should first be introduced into a small colony or divisions of one in a queen shipping cage. After she is laying, the small colony can be united with a large one.

A drone-laying queen can be replaced if she is discovered while the colony is still strong. If the colony is weak, the bees should be removed and the equipment added to another colony.

Assuming colony conditions and the condition of the queen are favorable, the effect of environmental or working conditions and the time of year are factors that affect queen acceptance. Best acceptance is usually obtained when some nectar is available in the field.

One possible period for requeening is during the broodless period of late fall. Queens are easily introduced at this time, and the bees are passive to their presence. However, the uncertainty of the weather, the difficulty of finding old and shrunken queens, and the danger of inciting robbing make this time of year less desirable for requeening than the summer.

Brood rearing declines in late summer and fall, and many normal colonies are completely broodless during much of November and December, particularly if the colony has no pollen. Older queens stop brood rearing sooner than younger queens.

Brood rearing should be encouraged as late in the season as possible. This can be assured by providing vigorous young queens in late summer, by preventing undue overcrowding and restriction of the brood nest with honey, and by encouraging pollen storage.

In areas where fall honeyflows occur, partially filled supers should be kept on the colonies, especially if the brood nest is heavy.

If brood rearing is restricted by a crowded brood nest or because of poor queens, the colony may enter the winter with a high precentage of old bees that will die early in the winter. Such colonies may later develop serious nosema infections and perish before spring. A colony should start the winter with about ten pounds of bees and plenty of honey to carry it to the next spring.

A

B

Queens with (A) irregular and (B) good brood pattern (USDA photograph)

Preparing Colony for Winter

POPULATION

The strength of a colony of bees is relative and difficult to describe. A "strong" colony to one beekeeper might be "weak" to another. Colonies with less than ten pounds of bees should be united to stronger ones or several weaker ones combined. At between 40° and 50° F., ten pounds of bees will cover practically all the combs of a three-story hive wall to wall and top to bottom. Naturally as the temperature drops the cluster will contract.

The beekeeper must see that at no time is the available space for brood rearing reduced because of overcrowding with honey from the fall flow. A balance must be maintained between crowding the colony to get the brood chambers well filled with honey and adding space to relieve brood rearing restriction. Partially filled supers kept on colonies in the fall may be necessary. Any subnormal colony should not be overwintered, but should be united with another colony.

A colony may appear to have an adequate fall population, but if the bees are old, it will weaken rapidly as winter advances and may starve to death, even with abundant honey in the hive, because the cluster is too small to cover the honey stores.

FOOD RESERVES

The colony should have a minimum of 500 square inches of comb filled with pollen in the fall. To insure uninterrupted brood rearing in late winter and early spring, the beekeeper may need to supplement this. The average colony of bees under intensive management may consume about sixty pounds of honey between the last fall flow and the first available food from the field in the spring. A weak colony may consume twenty pounds or less, but the very best colony will consume eighty pounds or more. To insure the survival of the top quality colony, ninety to 100 pounds of honey should be left on it in the fall. A colony of bees not rearing brood will consume an average of about one-eighth pound of honey per day or five pounds per month. When brood rearing begins, the consumption of honey is greatly accelerated. Brood rearing should commence in midwinter and accelerate as temperatures moderate in late winter and early spring.

When brood rearing is discouraged or curtailed, the colony will consume less winter stores but will emerge in the spring much weaker and with a population of primarily old bees. Such colonies will have difficulty replacing the small amount of honey they used over winter, whereas other colonies that have had normal, unimpeded rearing of brood will soon be able to replace all the honey they consumed over winter plus a substantial surplus.

ORGANIZATION

To accommodate the best queens in standard Langstroth ten-frame hives, a minimum of two hive bodies and preferably three should be used for year-round management. In the fall most of the honey should be located in the top hive body. With experience the beekeeper can soon learn to estimate the weight of hive bodies or frames by lifting them. A frame full of honey should weigh approximately five pounds. The top hive body should contain forty to forty-five pounds of honey. This means that all frames in the top hive body will be full of honey except for two or three frames in the center. The second body should contain twenty-five to thirty pounds of honey and some pollen. The bottom hive body should contain twenty to thirty pounds of honey plus pollen. If in the fall the combs in the top hive body are not filled, the beekeeper should reorganize them and if necessary feed additional sugar sirup so that this top hive body is well filled with stores.

As the winter progresses the cluster of bees will shift its position upward as the stores are consumed. A colony of bees in a cold climate can starve with abundant honey in the hive if the honey is below the cluster.

With the advent of cold weather, the bees cluster tightly in the interspaces of the combs. Usually there are no bees in the bottom part of the hive near the entrance. For this reason an entrance cleat or reducer should be used to exclude mice. One-inch auger holes drilled into the hive bodies of the brood nest just below the hand-holds are helpful. In late summer these auger-hole entrances are closed with corks so that the bees will fill the combs near them. During winter the top auger-hole entrance should be open. This allows the escape of moisture-laden air and affords a flight exit for the bees during warm spells.

PACKING THE HIVE

Many beekeepers in the coldest parts of the country consider that some form of protection around the hive is essential. Others believe that colonies with strong populations and ample stores need no further protection. Factors to consider in deciding whether or not to pack are the cost of material and labor and any savings in honey or bees. Packing will not replenish colonies deficient in honey, pollen, or bees, replace poor queens, or cure bee diseases. Packed colonies will consume slightly less honey. However, the difference is negligible. The most important consideration in preparing colonies for winter is a strong population and adequate stores.

Colony during winter showing top auger-hole entrance.

When outside temperatures are near freezing, the temperature at the surface of a cluster of bees ranges between 43° and 46° F. As the temperature decreases, the cluster contracts and the bees in the outer insulating shell concentrate to provide an insulating band of one to three inches in depth. Metabolism and activity of the bees in the center of the cluster maintain a desired temperature. This may be around 92° F. if brood rearing is in progress. The temperature of the area of the hive not occupied by bees will be similar to the external temperature. This is true whether the hive is packed or not. The difference is that the temperature in the unpacked hive changes more rapidly and responds more quickly to that outside the hive. Heavy packing is worse than no packing, because during warm periods in midwinter when the bees should fly, those heavily packed do not fly at all.

Late Winter Manipulation

If colonies are inspected in late winter or early spring, adjustments can be made to save colonies that might be lost otherwise. Even weak or medium-strength colonies can often be saved if honey is moved into contact with the cluster. A strong colony with insufficient honey can starve if additional food is not provided at this time.

From this period until the bees can forage, such colonies can be fed either full combs of honey, or if these are not available, a gallon or two of heavy sugar syrup (two parts sugar by volume to one part water) can be poured directly into the open cells of empty combs.

Spring Buildup

Overwintered colonies will usually start brood rearing in midwinter and continue into the summer unless the stored pollen is all consumed before fresh pollen is available. If the supply is exhausted and not supplemented, brood rearing will slow down or stop entirely when it should proceed without interruption.

For best results in honey production, a beekeeper should have strong populations of young bees for the honeyflow. Colonies emerging in the spring with predominantly old bees must build a population of young bees for later flows by using the early sources of pollen.

Some beekeepers trap pollen at the hive entrance from incoming bees by means of a pollen trap such as that described in U.S. Department of Agriculture ARS 33-111, *A Simplified Pollen Trap for Use on Colonies of Honeybees.* This pollen is dried or frozen until needed, then mixed with sugar, water, and soy flour, and fed to the colony as a supplement to its natural supply. Various other types of pollen supplements and substitutes have been described and some are available on the open market.

Supplements containing pollen are eaten more readily by bees and generally give better results than those containing soy flour or other

Strong colony feeding on pollen supplement cake during March. (USDA photograph).

material without pollen. Pollen supplement is preferred by the bees in direct proportion to the amount of pollen it contains. The less pollen the supplement contains, the less is eaten. Substitutes made without pollen tend to be dry and gummy. A pound of pollen will make approximately twelve pounds of pollen supplement.

Swarm Control

After pollen becomes abundantly available in the spring, the beekeeper should provide ample space for brood rearing and honey storage.

The natural colony behavior is to expand its brood nest upward, and a simple manipulation utilizing this tendency is to shift the empty frames or emerging brood to the top of the hive and the youngest brood and honey to the bottom part. This permits the expansion of the brood rearing upward into this area. Subsequent reversal of brood chambers can be made at about ten-day or two-week intervals until the honeyflow starts.

As soon as the three brood chambers are filled with bees, the first super should be given whether or not the honeyflow is in progress. If this is done, most colonies with a vigorous queen will not swarm. However, any queen cells the beekeeper sees as he reverses the brood chambers should be removed. A simple method of reversing brood chambers is to lower the hive backward to the ground, separate the brood chambers, interchange the first and third hive bodies, and return to position.

After the honeyflow starts, the danger of swarming lessens and brood chamber reversal can be discontinued. At the start of the honey-flow, "bottom supering" should be used. The empty super should be placed above the top brood chamber but below the partially filled supers.

After the supers have been filled and the honey has been extracted, they should never be put directly over the brood nest, but should be placed on top of the partially filled supers to prevent the queen seeking them and laying eggs in them. Why such combs are attractive to the queen is not known.

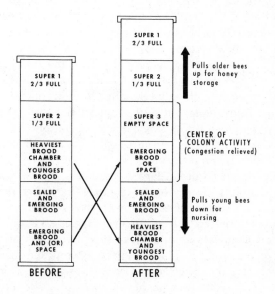

In three brood hive, this is basic manipulation to avoid swarming. (USDA photograph)

Two-Queen System

The establishment of a two-queen colony is based on the harmonious existence of two queens in a colony unit. Any system that insures egg production of two queens in a single colony for about two months prior to the honeyflow will boost honey production.

The population in a two-queen colony may be twice the population of a single-queen colony. Such a colony will produce more honey and produce it more efficiently than will two single-queen colonies. A two-queen colony usually enters winter with more pollen than a single-queen colony. As a result of this pollen reserve, the two-queen colony emerges in the spring with a larger population of young bees and is thus a more ideal unit for starting another two-queen system.

To operate two-queen colonies, start with strong overwintered colonies. Build them to maximum strength in early spring. Obtain young queens about two months before the major honeyflows start. When the queens arrive, temporarily divide the colony. Replace the old queen and most of the younger brood plus about half of the population in the bottom section. Cover with an inner cover or a thin board and close the escape hole. The division containing most of the sealed and emerging brood, the new queen, and the remainder of the population is placed above. The upper unit is provided with an exit hole for flight.

At least two brood chambers must be used for the bottom queen and two for the top queen. Two weeks after her introduction, remove the division board and replace it with a queen excluder. The supering is double that required for a single-queen operation. In other words, where three standard supers are necessary for a single colony, six will be required for a two-queen colony.

When supering is required, larger populations in two-queen colonies require considerably more room at one time than is required for single-queen colonies. If a single-queen colony receives one super, a two-queen system may require two or even three empty supers at one time.

The brood chambers should be reversed to allow normal upward expansion of the brood area about every seven to ten days until about four weeks before the expected end of the flow, after which the honey crop on the colony may be so heavy as to preclude any brood nest manipulations. Thereafter give supers as they are needed for storage of the crop. As the honey is extracted, the supers are returned to the hive to be refilled. They should never be replaced directly over the top

Brood combs showing (top) health brood necessary for high honey production and (bottom) diseased brood which results in weakened colonies and low honey production. (USDA photograph)

brood nest, unless a second queen excluder is used to keep the queen out of them. The top brood nest may tend to become honey bound. If this occurs, reverse the upper and lower brood nests around the queen excluder. This puts the top honey-bound brood nest on the bottom board and the lighter brood nest with the old queen above the excluder.

There is no advantage in having a second queen when about a month of honeyflow remains, because eggs laid from this time on will not develop into foragers before the flow has ended. However, entering the brood nest during the middle of the flow to remove one of the queens is impractical. Uniting back to a single-queen status can be done after the bulk of the honey is removed from the colony. By this time some of the colonies may have already disposed of one queen. When this happens, all that needs to be done is simply to remove the queen excluder and operate the colony as a single-queen unit.

Improved Stock

Production of honey is one major criterion in selecting honeybee stock and breeding for improvement. Superior stock must also be reasonably gentle, not prone to excessive swarming, maintain a large but compact brood nest, and winter well. It should ripen its honey rapidly, seal the cells with white wax, and use a minimum of burr comb. To obtain all the desirable characters in a superior stock, specific inbred lines from many sources must be selected and developed and then recombined into a genetically controlled hybrid. When this is done, hybrid vigor or heterosis usually results.

Queens of common stock reared under favorable conditions and heading well-managed colonies will probably be more productive than poorly reared queens of superior stock. Queens of superior stock reared under favorable conditions will require a higher standard of management than is demanded by common stock. To realize the maximum benefits from improved stock, the beekeeper must provide unrestricted room for brood rearing, ripening of nectar, and storage of honey.

To realize the maximum benefits from improved stock, the queen breeder should produce the best queens possible, and the honey producer receiving these queens should manage them in such a way that they can develop their maximum colony populations.

CHAPTER 14

Honey in Cooking

by Roger Griffith

Baked goods that stay fresh longer. New flavors in old dishes. A rich, brown color for breads and cakes. These are some of the benefits from using honey in cooking. You will find many more.

Caution is urged, however. Cooks describe honey as tricky to work with. One cup of sugar is just like another. Not so with honey. The taste varies from the light and mild to the dark and strong.

There is not one honey flavor, but many. The flavor depends on the flower from which the nectar was taken. Even honey from the same flower, such as the white clover, will taste different, depending on the area, condition of the soil and other factors.

The stronger honey, such as buckwheat, will provide not only sweetness but a distinct and sometimes overpowering flavor that the cook may or may not desire. Such honey, usually dark in color, is often preferred for cereal or toast. The lighter-colored honeys are recommended for first attempts at cooking.

Influencing these differences in flavor are the marked differences in sweetness and acidity of various honeys.

Honey's hydroscopic quality—its ability to absorb moisture from the air—also varies from one honey to another. It is this quality that makes baked goods made with honey remain fresh longer. It is also the quality that can make honey candy a sticky mess unless it is wrapped to prevent moisture from reaching it.

The cook should ease into the use of honey, experimenting to get the feel of it in recipes. An over-ambitious start with bad results could result in discouragement.

A common way to begin is to replace only one-half or one-third the sugar called for in a recipe. Many recipes for cakes and cookies call for both sugar and honey, with cooks maintaining they get a better texture by combining the two rather than using honey alone.

Here are some things to remember when cooking with honey:

1. A higher temperature is needed when using honey in making candy, jams and jellies.

2. Honey is sweeter than sugar, and of course has more moisture. One guide for baking is to substitute one cup of honey for each 1¼ cups of sugar, and, further, to reduce the liquid in the recipe by ¼ cup for each cup of honey used.

3. When substituting honey for sugar in cakes, add a quarter tea-spoon of baking soda for each cup of honey, to neutralize the acidity of the honey. If a cake does not rise properly, this indicates not enough baking soda was used. If the cake rises, but there is an unpleasant and darker layer at the bottom of the cake, too much baking soda was used. If the recipe calls for sour milk or sour cream, it is not necessary to add baking soda.

4. Freezing intensifies the honey flavor in baked products.

5. When measuring honey, use a greased cup or spoon, and the honey will flow freely and completely from it.

One tablespoon of honey equals sixty-four calories, slightly more than in granulated sugar. Honey contains minerals; its vitamin content is insignificant. Because it is a natural sweet, it is thought of as more wholesome than sugar. Persons eating it should know it will add pounds as quickly as sugar.

A cup of honey weighs twelve ounces; water, eight ounces; granulated sugar, seven ounces.

And here are some recipes, chosen to show how good honey is, and how wide is the range of its uses.

HONEY CORN MUFFINS

¾ cup sifted flour
1¼ teaspoons baking powder
½ teaspoon salt
⅓ cup cornmeal
1 egg, well beaten
⅓ cup milk
¼ cup honey
3 tablespoons shortening, melted
¼ cup pared diced apple.

Mix the flour, baking powder and salt. Stir in cornmeal. Combine egg, milk, honey, and shortening. Add all at once to cornmeal mixture; add apple; stir only enough to dampen flour. Spoon into eight well-greased, 2-inch muffin pans. Bake at 400° F. for 15 to 20 minutes.

HONEY CEREAL

3 cups rolled oats
1 cup wheat germ
1 cup sesame seeds
1 cup shredded coconut
¼ cup oil
¾ cup honey
1 teaspoon vanilla

Mix dry ingredients. Pour oil into measuring cup, tipping it to coat interior of cup. Add honey and vanilla and stir. Pour this over dry ingredients to coat them. Spread in shallow layer across cookie sheet. Stir occasionally while it is baking at 250° F. until it is golden brown. When cooled it can be stored in jars. Serve like dry cereal, with milk. This is a basic recipe, and each person should experiment with other combinations of ingredients until a favorite is found. Raisins, chopped nuts and chopped fruits may be added. And of course the taste improves if, when eating it, you add fresh strawberries or other fruits to your bowl.

STRAWBERRY SPECIAL

2 tablespoons honey
1 cup sliced strawberries
1 cup yogurt
1 cup milk

Your blender will do the rest, as you combine ingredients and whip until smooth. Hail the strawberry season with this.

HONEY GINGERBREAD

½ cup sugar
½ cup shortening
1 egg, beaten
1 cup honey
2½ cups sifted flour
1½ teaspoons soda
1 teaspoon cinnamon

Blend sugar and shortening, add egg and honey. Sift together dry ingredients and add. Pour in hot water and beat until smooth. Pour into two greased pans and bake at 350° F. for 35 minutes.

HONEY–PEANUT BUTTER COOKIES

¾ cup honey
1 egg
½ cup oil
½ cup peanut butter
1½ cups flour
½ teaspoon salt
1 teaspoon baking powder
1½ teaspoons orange juice
1 teaspoon vanilla

1 teaspoon cinnamon
1 teaspoon ginger
½ teaspoon cloves
½ teaspoon salt
1 cup hot water

Warm honey, mix oil and egg with it, then work in peanut butter. Add flour, baking powder, orange juice, salt and vanilla. Shape into small balls, flatten and bake on a greased cookie sheet for 10 minutes at 350° F.

HONEY-BAKED CHICKEN

12 pieces of chicken, thighs, legs and breasts
   oil
½ cup light honey
¼ cup soy sauce

Heat oil in large skillet, and brown chicken on all sides. Grease baking dish and arrange chicken in single layer. Mix honey and soy sauce and brush chicken pieces. While chicken is baking (200° F. for 2 hours), brush occasionally with honey mixture.

CARROTS FOR CHILDREN

18–24 small carrots, cooked.
   3 cups honey
   ½ cup water
   2 tablespoons powdered ginger
   2 tablespoons butter

Warm honey, add water, ginger and butter to it. Stir while cooking until it is thick, then pour over carrots and serve. This is a fine way to stop a child's plea that "I don't like carrots." For tastiest results, use thumb-sized carrots, possibly those being thinned in your garden.

HONEY NUT BREAD

1 cup honey
1 cup milk
¼ cup butter
2 beaten eggs
2½ cups whole wheat flour
1 teaspoon salt
1 tablespoon baking powder
½ cup chopped walnuts

Warm honey and combine with milk. When blended beat in remainder of ingredients, except for nuts. Fold them in when mixture is blended. Spoon into large greased loaf pan and bake for one hour at 325° F. This is an old recipe that is delicious. It freezes well.

HONEY BRUNCH COCOA

1 quart milk
2 sticks cinnamon
¼ cup cocoa
⅛ teaspoon salt
3 tablespoons honey

Scald milk with cinnamon sticks. Mix cocoa and salt; blend in ½ cup hot milk until smooth. Add to scalded milk and stir in honey. Remove the cinnamon sticks. Mix with a rotary beater. Serves 5.

CHAPTER 15

Queen and
Package Bee Production

by William C. Roberts and Warren Whitcomb, Jr.,
apiculturists, Entomology Research Division,
Agricultural Research Service

Rearing Queens

For the amateur who desires to rear a few queens, many methods are
available. Most of them are described in the references at the end of this
section. For those desiring to rear less than twenty queen cells at one
time and not more than fifty queens during a season, we suggest the
following procedure: Select the best colony as a breeder to produce the
queen cells. Rear them after the main honeyflow is over by two easy
manipulations.

First, remove the queen with one frame of brood and two frames of
honey with adhering bees to a hive body in another location. The
queenless colony will start a dozen or more queen cells within twenty-
four hours. These will be produced throughout the brood nest. Cut out
and save the largest and most uniform cell. Be sure that the cell selected
is not damaged. Destroy all unused cells. Return the queen and three

Reprinted in cooperation with Louisiana Agricultural Experiment Station from *Beekeeping in the United States*, USDA Agriculture Handbook 335.

frames to the center of the brood nest of the parent colony. If the bees are not robbing, there will be little chance of losing the queen by this method.

Finally, make an additional colony or nucleus by again removing from another colony three frames of brood, honey, and adhering bees. Gouge a cavity near the top of the center comb about ¾ inch wide and 1½ inches long. Gently press the queen cell into it so the queen will have room to emerge from the cell tip. Replace the comb and leave the new hive closed for two weeks. Repeat this process for each queen desired.

The commercial queen breeder does not use this inefficient method. He transfers female larvae from worker cells into artificial queen cell cups and uses special queenright or queenless cell-building colonies to rear them. By this method hundreds of cells can be produced continuously.

Commercial Queen Production

In the United States there are many commercial queen breeders who rear annually several thousand queens each. They are concerned with quantity production of high quality queens. High quality queens are necessary for repeat sales, and quantity production is necessary for profitable operations. Customer demand determines the kind or breed of queens produced. Each queen breeder uses methods that suit his own particular conditions.

Queen cells are started when worker larvae are transferred or grafted into wax cells and placed in special cell-starting colonies. Two queenless types are in common use. The "swarm box" is composed of approximately four pounds of bees confined in a box with combs of honey and pollen. The "queenless colony" is a hive with six pounds or more of bees and an open entrance permitting the bees to fly. In each of these types about fifty queen cells can be started. The cells are left in these starters for twenty-four to thirty-six hours and then transferred to the cell-finisher colonies. If the queen breeder uses the "double graft" method of producing queens, he removes the larvae from the started queen cell and replaces them with other young larvae that are allowed to remain.

Some cell-finisher colonies are queenless, others are queenright. They are populous colonies with ten pounds or more of bees and with both brood and bees of all ages. They must have ample food—both honey and pollen. The queenless cell-finisher requires the addition of brood and bees at regular intervals. The queenright cell-finisher has a brood nest and queen below the excluder. The queen cells are placed between frames of brood in the upper part of the colony. Started queen cells are added every three or four days. The frame containing the started queen cells is placed between the frames of unsealed brood.

A few queen breeders start and finish queen cells in the same queen-less colony. Others start and finish the cells in a queenright cell-builder colony.

Good cell-builders properly handled produce good queen cells and poor cell-builders produce poor queen cells. The developing queen larva feeds for only five days. If it is properly and adequately fed during this period, a good virgin queen results. If inadequately fed during this period, a poor queen results. The best queen rearing methods are those that provide for well-fed larvae during the five-day feeding period. Malnutrition results in small queens with fewer ovarian tubules.

Queen cells are removed from the cell-builder colonies nine or ten days after grafting. Ten-day old cells go directly to the small mating hive or nuclei, but nine-day cells are kept for one or two days in an incubator maintained at brood rearing temperatures. On the eleventh day after grafting, the virgin queen emerges from the queen cell. Emergence may be delayed by keeping the queen cells at a slightly lower temperature during the late pupal stage of the queen.

Many types and sizes of mating nuclei are in use. Most queen breeders use a three-frame nucleus containing a feeder. The frames vary in size from about three or four inches to deep Langstroth frames. Many are established with bees, brood, and honey. Others are started without brood or honey, and the feeder is filled with sugar sirup. The amount of bees needed to stock a nucleus varies from one-fourth to three-fourths pound. If brood is given, only one frame of brood is placed in each nucleus. Many nuclei are started with only two combs, allowing space for adding the bees to the nucleus box. Nuclei are usually closed for twelve hours if brood is given, but may be kept closed for two or three days if no brood is used. When only two frames are used in installing the bees in the nucleus, the third frame is added four days later. Many queen breeders make up nuclei in a building and confine them there for a day or two before moving them to the mating yard.

To insure proper mating, the queen breeder should have an adequate supply of mature drones at all times during the mating periods. Special drone-producing colonies should be maintained in the apiaries containing the nuclei. Producers who do not maintain plenty of drones usually have many inadequately mated queens.

Fourteen days after a queen cell is given to a nucleus, it should have a mated queen with eggs and young larvae. Such queens are ready for shipment and can be removed. Another queen cell can then be given. Many breeders prefer to wait until the next day after removal of a laying queen before giving another queen cell.

If the queen is to be shipped by mail, the queen cage should have candy and seven to ten attendant worker bees. If the queen is to be placed in a package of bees for shipment or if she is to be placed in a queen bank for storage, she should be caged without attendant bees and without queen cage candy.

A queen bank is a colony without a laying queen and sometimes without brood. Often 200 to 500 queens in cages are stored in one such colony. They can be kept in a queen bank for a month or longer if sufficient populations are maintained.

Some queen breeders prefer to ship large orders of queens in queen cage carriers rather than adding queen cage candy and attendants to each queen cage. A carrier consists of two Langstroth frames holding thirty-six queens in each frame, one frame of honey, and about 1½ pounds of bees. A three-frame standard nucleus box with screen top and bottom makes a satisfactory carrier package for shipment.

A commercial queen breeder will obtain 200 to 400 queen cells from each cell-building colony each month of operation. This is enough to supply queen cells for 100 to 200 nuclei. Few queen breeders average more than fifty mated queens for each 100 cells put into nuclei.

Improvements in the technique of artificial insemination of queen bees may soon eliminate the need of nuclei and possibly cut the cost of queen production. Ninety-five percent survival of queens mated artificially is not uncommon with present techniques.

The greatest customer demand for package bees and queens occurs in March, April, and May. Consequently, the large producers are located in areas with a climate that permits economical production before and during these months. Locations in the South and in California, with mild winters and early springs, are ideally suited for early production of populous colonies. The choice locations are those that have not only a favorable climate but also good floral sources of nectar and pollen.

An area without a good nectar flow can be satisfactory, but a good pollen flow is essential. Many producers feed sugar sirup to their colonies during February and March. Others find it necessary to feed continuously during their most active season. There is no substitute for an abundant pollen income during the spring and fall. Pollen storage in the fall is necessary for early spring buildup. Without adequate pollen sources after the middle of February the colonies will curtail brood production.

The queen and package bee production in the United States today is a $5 million industry. More than one million queens and over 500 tons of bees are produced and sold to beekeepers in the United States and Canada each year.

Package Bee Production

Commercial shipments of package bees for replacing or starting new colonies in the honey-producing regions were apparently first made in 1912 by H. C. Short from Fitzpatrick, Ala. The demand for package bees increased greatly after World War I and shipments of queens reached a peak about 1947. The Railway Express Agency probably handled 80 to 90 percent of all bee shipments during the 1920's and 1930's. Since then truck, air, and mail shipments have become increasingly important. Approximately 80 percent of all package bees are now trucked to northern honey producers.

The requirements for producing a maximum crop of bees and a maximum crop of honey are basically similar. Prolific queens, adequate equipment, abundant stores at the proper time, a knowledge of local nectar and pollen sources, and a system of management planned for the particular environment are necessary for both crops. Fundamental differences are (1) the length of time during which colonies must be stimulated for maximum brood production, (2) the amount of equipment used, and (3) the maintenance. The shipping season lasts for two to three months for the bee-producing colony as opposed to a honey-flow period of three to six weeks for the honey-producing colony. Three deep bodies are used for the bee-producing colony and a minimum of three deep bodies for the honey-producing colony. After each shaking, a population of about 18,000 bees—the level that gives the maximum production of brood per bee—is required for the bee-producing colony as opposed to a population of 60,000 or more bees for maximum honey production.

Methods of producing package bees have been carefully studied. A system of management that is successful in Baton Rouge, La., will be described and it can be easily adapted to other areas.

Preparing colonies for package bee production should begin in August, with a check of colonies for good queens and the replacement of failing queens. August and September can be a danger period in

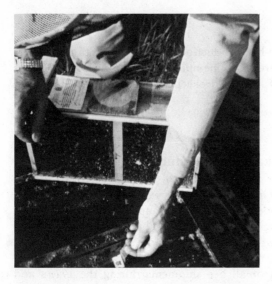

Moving bees from package to hive.

many parts of the shipping area because of lack of honeyflow—almost a necessity for successful queen introduction—and especially because of lack of pollen. Productive colonies will have mostly young bees in mid-November, although the colony may be nearly broodless. There should be fifty to seventy pounds of honey and 200 to 600 square inches of pollen in three deep bodies of good combs. About one-half the honey should be in the top body and the remainder fairly equally divided between the other two bodies of the brood nest. Three to six good combs of pollen should be placed near the center of the lower brood chambers. In good colonies most of this pollen will be used before the start of brood rearing in January. Lack of honey stores can be corrected by feeding sugar sirup. A deficiency in pollen can be partially connected by giving pollen cakes after brood rearing starts.

Colonies should reach maximum populations at the start of the shipping season. This means that colonies should be stimulated to sustained brood production for eight to ten weeks before colonies are to be shaken. The greatest production of brood per bee is in colonies with approximately 18,000 bees. The greatest colony yield of bees is obtained by shaking about four pounds of bees at regular ten-day intervals so that a population level of approximately 18,000 bees is maintained. At the start of the shipping season the strongest colonies may have 35,000 to 45,000 or more bees. Such colonies may have five pounds of bees shaken at ten-day intervals, but removing more than six pounds of bees at a single shaking results in decreased brood production.

Three-story colonies are necessary during the buildup period to supply the area for eggs and brood that a prolific queen requires together with the comb space necessary for honey and pollen stores. In order to decrease labor in handling hive bodies, most of the honey can be placed in the bottom body at the beginning of the shaking period and only the top two bodies manipulated. This three-story colony can be manipulated with essentially the same labor requirements as a two-story colony, but because of the space and food reserves in the bottom body the brood production will be higher. A maximum of thirty-seven pounds of bees has been shaken from a single colony and an average of thirty-two pounds of bees per colony from ten colonies during a sixty-day period without detriment to the colonies.

Methods of getting the bees into the shipping package vary widely depending on local honeyflow conditions, the type of package used, and how it will be transported. Two widely used methods include (1) shaking the bees from combs in the brood nest, which requires more labor but allows a constant check on the condition of the queen and colony, and (2) the smoke-up or drive-up method, by which bees are smoked or driven by a repellent up onto combs in the top body, then shaken off. By this method the brood nest combs are not removed. The recently developed forced-air method of removing bees from honey supers may be adapted for package bee removal. Bees are sometimes shaken through a funnel directly from the comb into the package, or into a "shaker box," from which they are later measured into packages. Both of these methods are satisfactory and widely modified by individual shippers.

A recent survey of commercial package bee shippers shows that most of them ship about ten pounds of bees from each colony during the shipping season.

CHAPTER 16

Identification and Control
of Honeybee Diseases

Prepared by H. Shimanuki, *Northeastern Region,*
Agricultural Research Service

Bee diseases are present throughout the United States. They are responsible for large annual losses in bees, honey, and equipment, and
add greatly to the cost of production. Also, the loss of pollinating bees
results in a lower yield of seed and fruit.

Bee diseases should be detected in their early stages; prompt treatment will prevent their spread. (See next page on sending samples
for diagnosis.) Contagious diseases spread quickly within a colony, and
the crowding of colonies increases the possibility of the spread of diseases from hive to hive. When searching for disease symptoms be
aware that a colony may have more than one disease.

It is especially important that the two most serious brood diseases—
American foulbrood and European foulbrood—be detected early.
Make routine inspections for these diseases.

Reprinted from *Identification and Control of Honeybee Diseases,* USDA Farmers'
Bulletin 2255.

Brood Diseases

To identify brood diseases, carefully examine dead brood found in the cells. The appearance of the combs may indicate which brood disease is present, but final diagnosis depends on the symptoms shown by the dead brood.

Dead brood in open cells of a comb can be seen clearly if the comb is inclined so that direct sunlight falls ·on the lower side of the cells. If you do not find any dead brood in the open cells, remove sunken, discolored, or punctured cappings and examine them for dead brood.

When you examine dead brood, observe its appearance and position in the cells. Note its age, color, consistency, and odor. For example, if the affected brood is unsealed in the comb then European foulbrood is suspected. If only the sealed brood is affected, and has collapsed into a ropy brown mass, American foulbrood is suspected.

SENDING DISEASED SAMPLES FOR DIAGNOSIS

It is sometimes difficult to make a definite diagnosis of diseased brood and honeybees in the apiary. Diagnosis in the laboratory is a service made available to beekeepers and State apiary inspectors by the U.S. Department of Agriculture.

When you select a sample of the comb for laboratory examination for brood diseases, cut a four-inch-square section of the comb. Make sure this piece of comb contains as much of the dead or discolored brood as possible.

If you suspect an adult bee disease, send at least 200 sick or recently dead bees in the sample. Pack this sample in a wood or strong cardboard box. Do *not* pack samples in tin or glass containers, and do not wrap either the comb or bees in waxed paper or aluminum foil.

Send all samples to the Bio-environmental Bee Laboratory, Agricultural Research Center, Beltsville, MD 20705. Print or type your name and address on the return label and include your Zip Code number.

AMERICAN FOULBROOD

Cause. American foulbrood, the most widespread and most destructive of the brood diseases in the United States, is caused by a spore-forming germ known as *Bacillus larvae*. Adult bees are not affected by this disease.

Only the spore stage of *Bacillus larvae* is infectious to honeybees. All castes of honeybees are susceptible to the disease, but worker larvae are particularly susceptible. However, larvae become immune to the disease about three days after the eggs hatch.

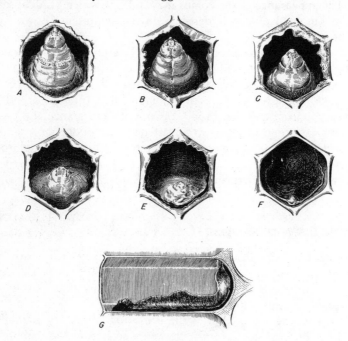

Honey bee larvae killed by American foulbrood, as seen in cells: (A) Healthy larva at age when most of brood dies of American foulbrood: (B-F) Dead larvae in progressive stages of decomposition. Remains shown in F are scale. (G) Longitudinal view of scale. (USDA photograph)

Effect. Only a few dead larvae or pupae will be seen when the colony is first infected by the disease. Occasionally, enough larvae become infected to weaken or kill the colony the first season. On the other hand, the disease may not develop to the critical stage until the following year. If left unchecked, American foulbrood quickly spreads to healthy colonies in nearby apiaries.

Symptoms. If American foulbrood is infecting your colony, the cells of your brood comb will have a scattered and irregular pattern of capped and uncapped cells, and cells with sunken and punctured cappings. This "pepperbox" appearance will contrast with the entirely sealed cells of a healthy brood comb.

A larval color change is one of the signs of infected larvae. Dying larvae gradually change from pearly white to dark brown.

The "pepperbox" pattern begins to form as the larvae shrink; the capping is drawn down into the cell so that the normally convex capping becomes concave. In advanced stages of the disease many of the cappings are punctured. (Cappings over dead brood are often removed by adult bees.)

The decay and drying of dead brood takes a month or more. The larval remains are dry and later form a brittle scale that adheres to the lower wall of the cell.

One way to verify that American foulbrood is causing larval death is to remove and examine some of the decayed larvae from the brood cells. During the early stages of decay, about three weeks after death, the body wall of the cell can be easily ruptured. Using a toothpick or matchstick, thrust into the decayed larvae and withdraw the decaying mass. If the disease is present, an inch or more of brown, gluelike thread can be withdrawn. This condition is known as the "ropy stage."

When the remains begin to turn brown and become ropy, an odor can be detected that is typical of the advanced stages of this disease. This same odor persists even when the scales are formed.

Larvae often develop into pupae before death occurs. Pupae undergo the same changes in color and consistency as larvae.

When a pupa dies from American foulbrood its tongue generally protrudes from the scale to the center of the cell. This symptom is characteristic only of pupae infected by American foulbrood. It should

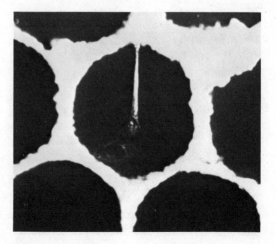

Pupal tongue from honeybee pupae killed by American foulbrood protrudes from the scale to the center of the cell. (USDA photograph)

not be confused with the short blunt protrusion—or "false tongue"—associated with European foulbrood.

Spread. American foulbrood can spread in a colony when:

- Nurse bees transmit bacillus spores to young larvae.

- Honey is stored in cells that once contained diseased brood.

- Bees are exposed to contaminated honey.

- Equipment is used for both diseased and healthy colonies.

Nurse bees can inadvertently feed bacillus spores to young larvae. Soon after the larva has been sealed in its cell, or just after it changes to a pupa, the spores will germinate in the gut of the larva and multiply rapidly, causing death. New spores will form by the time the larvae die. When the house bees clean out the cell containing the dead larva, spores will be distributed throughout the hive, thus infecting more larvae.

Honey stored in cells that once contained diseased brood becomes contaminated and may be fed to susceptible larvae. As the infection weakens a colony, the colony cannot defend itself from robber bees from strong colonies. The robber bees take the contaminated honey to their own colony and repeat the cycle of infection and robbing.

When bees are exposed to contaminated honey, or the same equipment is used for diseased and healthy colonies, there is a danger of disease spread. Therefore, it is extremely important that diseases are detected in their early stages, and that equipment is free from disease organisms.

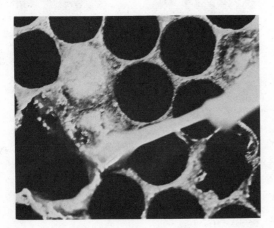

One way to determine whether American foulbrood is causing larval death is to remove and examine decayed larvae from the brood cells. Here, removal of a gluelike thread, known as the "ropy stage," proves the disease is present. (USDA photograph)

Control. No commonly used strains of honeybees are immune to American foulbrood.

Burning. When American foulbrood is discovered in your apiary the diseased colonies should be destroyed by burning. Before you burn diseased colonies, however, you must dig a pit to hold the burned material.

Dig a pit eighteen inches or more in depth and wide enough to hold all the material to be burned. Situate the pit in a place not likely to be disturbed.

To kill the bees, place a tablespoon of calcium cyanide in the entrance of the beehive on a sheet of paper or cardboard. Death will occur within minutes.

Be sure all flight activity has ceased before applying calcium cyanide. Calcium cyanide is poisonous and should be handled with extreme care. To allow the fumes to penetrate to all parts of the hive, place additional cyanide on a piece of paper above the top bars, and under the inverted inner cover. Then replace the hive cover.

Immediately after all the bees have been killed, place the hive on pieces of burlap or strong paper; this will make it easier to gather up and burn the bits of comb, honey, or dead bees. Do this quickly to reduce the possibility of robber bees spreading the disease to healthy colonies.

When burning the colony, kindle a fire beneath it with cross members strong enough to support the weight of the frames. Allow plenty of ventilation. A brisk, hot fire is necessary to quickly burn the brood and honey.

Do not burn the bottom boards, hive bodies, inner covers, or outer covers. These items should be scraped to remove all bee glue (propolis) and wax, and then scrubbed with a stiff brush and hot soap solution. Afterward, dispose of the wash water and burn the scrapings so they are not accessible to bees.

The scraping and scrubbing procedures above *will not* sterilize the bee equipment. To do this, completely immerse your equipment for twenty minutes in a boiling lye solution (sodium hydroxide) containing one pound of lye to ten gallons of water. Wooden parts can be damaged by longer exposure. Weaker solutions may not remove all the wax and propolis from the hive equipment.

Remember that lye solutions are caustic and can cause severe burns. Before using the lye, read the label carefully and observe all precautions.

Treatment. The burning of diseased colonies in the apiary gets rid of only those colonies in which American foulbrood is present in an active form. To prevent the spreading of the disease throughout the apiary, and to control it, use oxytetracycline (Terramycin) or sodium sulfathiazole. The U.S. Food and Drug Administration has permitted the use of both these drugs.

Labeling for these drugs carry specific instructions for the application of both these materials, subject to state laws and regulations. Consult with state apiary inspectors, extension apiculturists, or state entomologists before using these or any chemicals. Take care not to feed the colony oxytetracycline or sodium sulfathiazole in a manner that will contaminate marketable honey.

WARNING

Do not feed any medication to your colony when there is any danger of contaminating the honey crop. Be sure to stop all drug feeding at least four weeks before the surplus honeyflow.

After you have completed treatment with these drugs, remove all traces of honey from the frames and from any surplus honey combs to avoid contamination of honey to be used for human food. If additional drug treatment is required during surplus honeyflow, the honey must not be used for food. Honey from bee colonies infected by American foulbrood should not be used for preparation of medicated sirup. The sirup may be contaminated with disease-causing organisms.

Both sodium sulfathiazole and oxytetracycline will prevent American foulbrood, but they do not destroy the spores of *Bacillus larvae.* Virulent spores may still be present in the colonies. Therefore, it is necessary to keep treated colonies under close observation.

EUROPEAN FOULBROOD

Cause. European foulbrood is caused by the germ *Streptococcus pluton.* These lancet-shaped bacteria are usually present in large numbers in sick and recently dead larvae.

European foulbrood is found all over the world; in some areas it is considered far more serious than American foulbrood.

Effect. All castes of bees are susceptible to European foulbrood, although differences in susceptibility can be found in various stocks.

European foulbrood is most common in the late spring. This is a period when brood rearing is at its height, although the earliest brood is rarely affected.

Sometimes the onset of the disease is quite subtle and difficult to detect. It spreads slowly through the colony with little apparent damage. In severe cases, however, colonies are seriously weakened.

The disease usually subsides by midsummer, but it occasionally stays active through summer and fall. Sometimes the disease subsides in the summer and reappears in the fall. A good honeyflow hastens recovery.

Symptoms. In European foulbrood the "pepperbox" pattern of capped and uncapped cells develops only when the disease attains serious proportions. Unlike American foulbrood, most of the larvae die before their cells are capped. However, you can sometimes observe discolored, sunken, or punctured cappings.

The most significant symptom of European foulbrood is the color change of the larvae. They change from normal glistening white to a

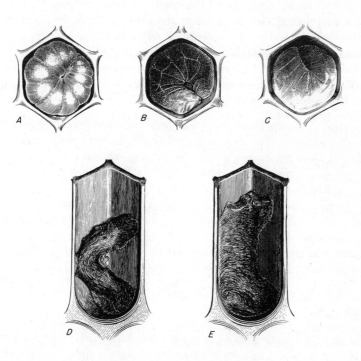

Honeybee larvae killed by European foulbrood, as seen in the cells. (A) Healthy larva at earliest age when the brood dies of European foulbrood. (B) Scale formed by a dried-down larva. (C) One of several positions of sick larvae prior to death. (D) and (E) Longitudinal views of scales from larvae that assumed a lengthwise position prior to death. (USDA photograph)

faint yellow. Larvae also lose their plump appearance and appear undernourished.

Most larvae die while in the coiled stage. When the larvae become brown their white tracheal system becomes visible. The diseased larvae sometimes appear to collapse from their upright state. In such cases larval remains appear twisted or melted to the bottom of the cell.

Recently dead larvae are rarely ropy. Scales can be removed easily from the cells and are rubbery rather than brittle as in American foulbrood.

The odor of European foulbrood varies. Typically, as sour odor is present from decayed larvae.

Spread. Spores are not formed by *Streptococcus pluton* but the organism often overwinters on combs. It gains entry into the larvae in contaminated brood food, multiplies rapidly within the gut of the larvae, and causes death about four days after egg hatch.

Not all infected larvae die from the disease. Sometimes larvae develop normally and void the germ, or regurgitate the bacteria onto the underside of the capping. These materials become sources of the disease.

Since the honey of infected colonies is contaminated the disease can be spread by robber bees or by the interchange of contaminated equipment among colonies and drifting bees.

Control. In some cases, European foulbrood can be eliminated by requeening colonies with a young queen. Requeening accomplishes two things: it gives the colony a more prolific queen, and it permits a time lag between brood cycles that allows the house bees to remove diseased larvae from their cells.

Treatment. The antibiotic oxytetracycline is approved by the U.S. Food and Drug Administration for the prevention and control of European foulbrood. Consult with state apiary inspectors, extension apiculturists, or state entomologists before using this drug. Follow the same precautions called for in the control of American foulbrood.

SACBROOD

Cause. Sacbrood is caused by a filterable virus. This virus is so small that it cannot be seen even with the aid of a light microscope.

Effect. Sacbrood rarely wipes out a colony of bees or becomes a serious menace to beekeeping. The disease affects both workers and drones. Pupae are killed occasionally, but adult bees are immune.

It is important for you to recognize sacbrood and distinguish it from the more serious foulbrood diseases. For a comparison of brood disease characteristics see the guide below.

Sacbrood is most common during the first half of the brood-rearing season. The disease often goes unnoticed because it affects only a small percentage of brood.

Symptoms. In sacbrood, death usually occurs after the cells are first sealed. The disease rarely reaches the serious stage in which the "pepperbox" pattern becomes evident.

The larvae gradually change from pearly white to dull yellow or gray, and finally to black.

The head of the larva is the first part of the body to change color. It becomes black. The larva dies in an upright position.

When you remove diseased larvae from their cells and examine them, you will observe that the contents of the larvae are watery and the skin is tough and forms a sac. Hence the name: sacbrood.

The scale is gondola shaped; both the head region and posterior bend toward the center. Sacbrood scales are rough and brittle, and they loosely adhere to the cell walls.

Spread. Very little is known about the transmission of sacbrood. However, experiments employing a suspension made from diseased larvae indicate that the disease can be spread to healthy larvae through contamination of larval food.

Nurse bees are suspected of transmitting the disease by carrying the virus from cell to cell. It is also believed that robber bees spread the disease by carrying contaminated honey from colony to colony.

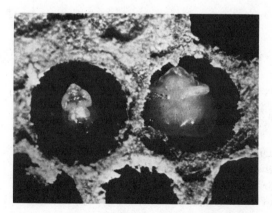

In sacbrood, the larva dies in an upright position. Left, the erect head of a diseased larva. Right, the head of a healthy larva. (USDA photograph)

Control. No chemotherapeutic agent is effective in preventing or controlling sacbrood. Requeening of colonies gives some degree of success. Generally, a colony will recover from sacbrood without a beekeeper's aid.

Fungus Diseases

Relatively little is known about bee diseases caused by fungi. The two most significant fungus diseases are chalkbrood and stonebrood.

Chalkbrood is caused by the fungus *Ascosphaera apis.* In 1968, the first case of chalkbrood was reported in the United States. Before then, it was solely a European problem. The disease has been reported in California, Minnesota, Montana, and North Dakota.

Young pupae or recently sealed larvae are the most susceptible to chalkbrood. Diseased bees are usually covered with filaments that have a fluffy, cottonlike appearance. If the fungus produces spores, the diseased brood will turn gray or, sometimes, black.

Chalkbrood is seldom considered serious because it is not highly contagious. Colonies eliminate this disease without beekeeper help.

Stonebrood is caused by the fungi belonging to the genus *Aspergillus,* primarily *A. flavus.* These fungi can be found in soil, on accumulations of dead bees, and on honeycombs.

When larvae and pupae are infected by this disease they have a green, powdery substance on their bodies. Spores form earliest and most abundantly near the head end of the dead larvae.

After dead larvae and pupae become dry, they are known as mummies. The disease is called "stonebrood" because of the hard texture of the dead brood.

Infection With Two or More Diseases

When more than one disease affects a comb it is easy to overlook a mixed infection. For example, larvae infected by American foulbrood have been found in the same comb with larvae infected by European foulbrood or sacbrood.

No single larva, however, has been found to be infected with more than one disease. Even when another disease is known to be present, make a careful search for American foulbrood. This is important when disease samples are sent for diagnosis—be sure that the sample is representative of the diseased comb.

Non-Infectious Diseases

Many beekeepers fail to recognize non-infectious diseases. Often, they assume that discolored larvae or pupae are the result of some infectious agent.

GUIDE FOR COMPARING CHARACTERISTICS
OF VARIOUS BROOD DISEASES OF HONEYBEES

Characteristics to observe	American foulbrood	European foulbrood	Sacbrood
Appearance of brood comb.	Sealed brood. Discolored, sunken, or punctured cappings.	Unsealed brood. Some sealed brood in advanced cases with discolored, sunken, or punctured cappings.	Sealed brood. Scattered cells with punctured cappings, often with two holes.
Age of dead brood.	Usually older sealed larvae or young pupae.	Usually young unsealed larvae; occasionally older sealed larvae.	Usually older sealed larvae; occasionally young unsealed larvae.
Color of dead brood.	Dull white, becoming light brown, coffee brown to dark brown, or almost black.	Dull white, becoming yellowish white to brown, dark brown, or almost black.	Grayish or straw-colored becoming brown, grayish black, or black; head end darker.
Consistency of dead brood.	Soft, becoming sticky to ropy.	Watery to pasty; rarely sticky or ropy.	Watery and granular; tough skin forms a sac.
Odor of dead brood.	Slight to pronounced glue odor to glue-pot odor.	Slightly to penetratingly sour.	None to slightly sour.
Scale characteristics.	Uniformly lies flat on lower side of cell. Adheres tightly to cell wall. Fine, threadlike tongue of dead pupae adheres to roof of cell. Head lies flat.	Usually twisted in cell. Does not adhere tightly to cell wall. Rubbery.	Head prominently curled up. Does not adhere tightly to cell wall. Lies flat on lower side of cell. Rough texture. Brittle.

In the spring, dead adult bees and some larvae or young pupae are found at the entrance to the hives. This is not always indicative of a contagious disease; brood is often neglected because of a shortage of nurse bees and then dies from either chilling or starvation.

Typically, the center of the brood cells appears normal—only those larvae on the periphery of the combs appear abnormal. This colony condition could be caused by an infectious disease of adult bees or by toxic chemicals and is not necessarily a brood disease.

Poisonous plants can cause non-infectious diseases. Purple brood is a common disease of this type in the Southeastern United States. It is caused by either the pollen or nectar from the southern leatherwood, *Cyrilla racemiflora L.*

Adult Bee Diseases

Adult bee diseases are not a threat in this country, except for nosema disease. In some foreign countries, acarine disease poses a potential threat to beekeeping.

NOSEMA DISEASE

Cause. Nosema disease is caused by the protozoan *Nosema apis.* The spores of the disease are ingested by the adult bee where they germinate and multiply in the gut.

Effect. Nosema disease decreases the effective life span of adult workers and, therefore, results in a decrease in the honey harvest. Infected queens are often replaced if their egg laying capacity is affected by nosema disease. Losses in the adult population can indirectly result in neglected brood.

Nosema disease is most prevalent in the spring. This is a time when a dysenteric condition tends to be present in the colonies. In such cases nosema disease poses a serious threat. Mixed infections of *Nosema apis* and the amoeba organism, *Malpighamoeba mellificae,* are believed to be more serious than either parasite alone.

Symptoms. A microscope is required to detect the spores of *Nosema apis.* The best time for finding spores of the disease is the beginning of the flying season after winter confinement. During the summer months it is difficult to find spores in the bees. A small increase of bees infected by the disease may be found in the fall.

At top is the digestive tract from a healthy bee. Note the individual circular constrictions on the ventriculus, or stomach, which is the tube at left. At bottom is a digestive tract of a honey bee with Nosema disease. Note that the circular constrictions on the ventriculus are not clearly defined. (USDA photograph)

No single symptom typifies this disease. Affected bees make trembling movements and their wings become unhooked. Unable to fly, these bees crawl about at the entrance to the colony.

The abdomen of an infected bee is often distended and appears shiny. The individual circular constrictions of a healthy bee's midintestine are visible, but in infected bees the midintestine may be swollen and the constrictions not clearly visible.

Spread. Nosema disease can be transmitted in several ways. In overwintered colonies, confined bees infected with nosema may defecate on the combs, causing healthy bees to become infected as they clean the combs in the spring. Food stores and soiled shipping cages can also be sources of infection.

Control. You can arrest the infection by giving the bees a supply of sugar sirup containing the antibiotic fumagillin. Because this drug has no effect on the spores of the nosema parasite this treatment will not completely eliminate the disease from the colony. The infection will continue when all the medicated sirup has been consumed.

Fumagillin is the only drug approved by the Food and Drug Administration for the prevention and control of nosema disease. Fumagillin is available on the market as *Fumidil-B®* .

No medication should be fed to the colonies when there is danger of contaminating the honey crop. The time of medication varies with location of the apiary.

Treatment. To decontaminate soiled bee equipment heat the equipment at 120° F. for twenty-four hours. This treatment will either destroy the spores or make them nonviable.

Take the following precautions when you give your equipment a heat treatment:

- Examine the combs beforehand and make sure they contain little or no honey or pollen.

- Stand the combs up so they are heated in their normal upright position (not on their sides).

- Check to see that boxes and combs have space between them that allows for air circulation.

- Never allow the temperature to exceed 120° F. or the wax will melt. Circulate the air in large rooms so no hot spots are created.

Acetic acid fumigation can also be used to decontaminate bee equipment. When giving this treatment stack hive bodies containing combs on a floorboard outdoors or in an open shed. Place a pad of cotton or other absorbent material previously soaked in ¼ pint of acetic acid (80 percent) on the top bars of the comb in each hive. Block the entrance and cover the stack with a wooden top. Seal any cracks in or between the hive bodies with masking tape. Leave the stack undisturbed for one week. After fumigation, air the combs for about forty-eight hours before using them again.

PARALYSIS

Cause. Paralysis of adult honeybees is a condition brought about by filterable viruses and by poisonous plants. The "paralysis" discussed here refers to the disease caused by the chronic, bee-paralysis virus.

Effect. Colonies can be affected by paralysis during the entire bee season. However, paralysis is more commonly found in warm climates. The disease affects only a small percentage of the bees. In severe cases, honey production of a colony can be seriously reduced by this virus. It is rare for colonies to be destroyed by paralysis disease.

Symptoms. Bees affected by the disease are usually found on the top bars of the combs. Individual bees tremble uncontrollably and are unable to fly. (Sick bees are sometimes attacked by healthy bees. When this condition is serious, large numbers of bees can be found crawling out of the colony entrance.) Bees with this condition are often hairless and have no control of their wings and legs. Abdomens of affected bees may be dark, shiny, or greasy.

Spread. How the virus is transmitted from bee to bee is not known. The paralysis virus is endemic in some colonies and the disease recurs each year in a small percentage of the bee population.

Control.* No chemical agent for the control of paralysis is available. Infected colonies seem able to cope with the disease without medication. The offspring of some queens appear to be more susceptible than others. Consequently, requeening of affected colonies is often effective in eliminating paralysis.

SEPTICEMIA

Septicemia is rarely considered a serious disease. Little is known about the bacterium, *Pseudomonas apiseptica,* that infects the bees.

Bees that die from septicemia often have a putrid odor. The muscles of the thorax decay rapidly and the body, legs, wings, and antennae fall apart when handled.

No control measure is known for this disease. The colony usually recovers spontaneously from septicemia.

AMOEBA DISEASE

Amoeba disease is rarely found in honeybee colonies. Losses from this disease are minor except when it is found in combination with other adult diseases.

---

*Mention of a proprietary product in this publication does not constitute a guarantee or warranty of the product by the U.S. Department of Agriculture and does not imply its approval by the Department to the exclusion of other products that may also be suitable.

Diagnosis of this disease requires examination with a microscope. Only the cyst stage of the pathogen *Malphighamoeba mellificae* is known. The disease is believed to be transmitted through the feces.

There is no known treatment for amoeba disease. Colonies are seldom if ever destroyed by this disease.

BEE-LOUSE

Braula coeca, commonly known as the "bee-louse," is not a louse, but a highly specialized parasitic fly, belonging to the order Diptera. *Braula coeca* adults are wingless and incapable of flight.

The tunnels made by the bee-louse's larvae in the honey combs destroy the market value of the comb.

The bee-louse can usually be found singly or in great numbers on the bee's thorax and mouth. It literally takes food out of the mouth of the honeybee.

The bee-louse is found in limited geographical areas. It has been seen in Maryland and Virginia. No effective treatment is available for infestation by the bee-louse.

ACARINE DISEASE

Acarine disease is not present in the United States. This disease causes serious losses of adult bees in Europe, and to some degree in South America and India. *For this reason, the importation of live adult honeybees into the United States, except from Canada, is prohibited by federal law.* No chemical agent is approved by the Food and Drug Administration for use against acarine disease.

Preventing the Spread of Bee Diseases

Bee diseases are spread when bees rob a diseased colony. For this reason good management requires that you minimize the opportunity to rob. Bee glue (propolis) and burr combs should be placed in containers which prevent the access of bees to the material. When a colony dies, close the hive to prevent the remaining stores from being robbed.

In most states, sale of equipment from colonies infected by American foulbrood is prohibited by law. Before you purchase any used equipment, be sure to consult your apiary inspector for information on the source of the equipment. As an added precaution, disinfect used equipment before use.

If you obtain adult bees or brood or feed honey (whether extracted or in combs) from unknown sources, do not add them to healthy colonies. Be certain that your source of bees and honey is from disease-free colonies. This is especially important when capturing swarms of unknown origin. When in doubt, isolate the colony until you are certain that it is disease free.

Inspect your bee colonies often. Watch for signs of disease. If any colony shows symptoms that are suspicious call your apiary inspector for his assistance, or send a test sample to a state or federal laboratory.

CHAPTER 17

Controlling
The Greater Wax Moth

by Agricultural Research Service

The greater wax moth[1] is also known as the bee moth, the bee miller, the wax miller, and the webworm. In its larval stage it damages combs and honey and is responsible for large losses to beekeepers in the United States. This insect is found almost everywhere that bees are raised. Its greatest damage is done in the Southern States, where its season of activity is longest.

Nature of Damage

The greater wax moth is most destructive to combs in storage, especially to combs stored in dark, warm, poorly ventilated places. The larvae of the moth tunnel into the combs, leaving them a mass of webs and debris.

Greater wax moths sometimes attack combs within the active hive, though such attacks are less common than those on stored combs. If the colony is strong, the bees defend themselves well against attack, and chance of infestation is slight. However, weak, diseased, starved,

---

[1] *Galleria mellonella*
Reprinted from *Farmers Bulletin No. 2217*, U.S. Dept. of Agriculture.

Adult female of the greater wax moth. (USDA photograph)

or otherwise abnormal colonies are a prey of the greater wax moth, and in these colonies the combs are often destroyed. Thus, though greater wax moths may not destroy a healthy colony, they may con-tribute to the destruction of an already weakened colony.

The larvae of the greater wax moth also do considerable damage to comb honey. The eggs are probably laid on the comb or section boxes before the comb-honey supers are removed from the hives, but the damage does not become evident until some time after the honey has been placed in storage. The damage consists of small, rather inconspicu-ous tunnels and borings made by the larvae through the thin wax caps of the honey cells. The honey leaks out through these holes, making the affected section unmarketable.

Description and Development

The greater wax moth passes through three stages of development be-fore becoming an adult—egg, larva, and pupa. In the Southern States these stages are not confined to particular times of the year. All stages may be present at any time during the year, and development is con-tinuous except during periods of low temperature.

THE EGG

The egg of the greater wax moth is small, white, and slightly oblong; its greatest diameter is less than one-fiftieth of an inch. Normally the female lays eggs in masses rather than singly, but even these masses are usually very difficult to see.

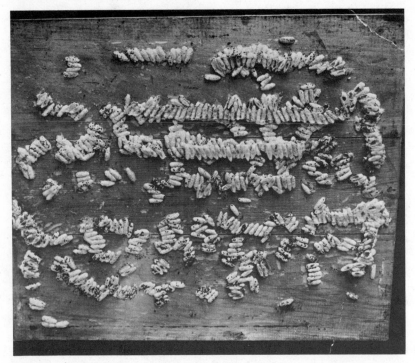

Cocoons of the greater wax moth. (USDA photograph)

The eggs are laid most frequently in the cracks between hive parts—that is, between supers, between hive body and bottom board, or between top super and cover. The egg masses may be deposited in these cracks from outside the hive or, if the colony is weak, from inside the hive. Egg laying within the hive almost always occurs in places farthest from the light.

At 75° to 80° F. the eggs hatch in five to eight days, but at lower temperatures (50° to 60° F.) the incubation period may extend to thirty-five days.

THE LARVA

The newly hatched larvae are often seen on the inner covers of hives and in cracks between supers and hive parts. They are white, extremely small, and very active. Almost immediately after hatching they attempt to burrow into the wax. The first attempts often do little more than roughen the surface of the wax; but after the first day, the larvae tunnel into the cell wall and make their way toward the midrib of the comb.

The length of the larval period ranges from twenty-eight days to nearly five months. During this period the larvae grow from about $1/25$ inch to as much as $7/8$ inch in length. The rate of growth and final size of the larvae depend chiefly on the food supply and temperature.

The larvae receive most of their nourishment from impurities in the wax, and in obtaining these impurities they ingest the wax itself. Foundation, which contains less of the impurities than the darker brood combs, is seldom attacked. Small larvae can develop on foundation, but many of them die, and those that survive develop at a relatively slow rate.

It is almost certain that some of the damage attributed to the lesser wax moth *(Achroia grisella)* is the work of these poorly fed greater wax moth larvae *(Galleria mellonella)*.

Temperatures most favorable for development of the larvae are between 85° and 95° F.—about those normally found in a beehive during the active season. At lower temperatures growth is retarded; at 40° to 45° F. no feeding or growth takes place, and the larvae seem to become dormant.

Part of a comb damaged by the webs and tunnels of greater wax moth larvae. (USDA photograph)

THE COCOON

When fully grown the larva spins a dense, rough silken cocoon. Some cocoons are found amid the tunnels and webbing in the combs, or in the refuse on the bottom of the hive; but usually the cocoon is firmly attached to some solid support, such as the frame, the side of the hive, or the inner cover. Frequently the larva cements its cocoon inside a cavity that it has chewed in the wood. These cavities sometimes extend completely through the end or top bars of the frame.

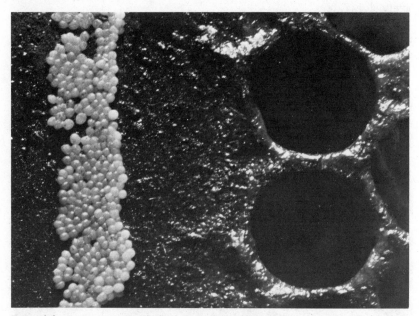

Eggs of the greater wax moth, laid on a comb. (USDA photograph)

THE PUPA

Within the cocoon the larva changes to the pupa. The duration of the pupal stage within the cocoon ranges from eight to sixty-two days; the higher the temperature, the shorter the duration. As with many other insects, the pupal period allows the greater wax moth to pass through the fall and winter protected against harmful weather conditions. In the South, especially in warm winters, the adults may emerge at any time.

Larvae of the greater wax moth. Left, dorsal view. Right, lateral view. (USDA photograph)

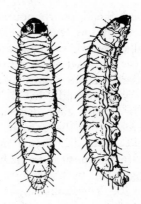

THE ADULT

The normal adult of the greater wax moth is about ¾ inch long and has a wingspread of 1 to 1¼ inches. The males are slightly smaller than the females and may be distinguished from them by the shape of the outer margin of the fore wing, which is scalloped in the male but smooth in the female. Adults are commonly seen in the resting position with their grayish-brown wings folded in rooflike fashion. The moths are not easily disturbed, but when molested they run rapidly before they take wing.

The moths vary widely in size and color, according to the type of food consumed by the larvae and the length of the developmental

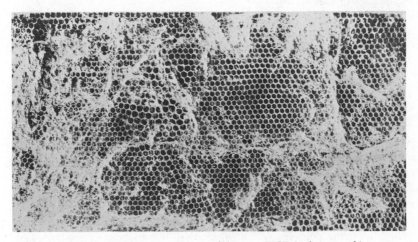

Brood comb infested with greater wax moth larvae. (USDA photograph)

period. Small, poorly nourished larvae, or those whose development is slowed by low temperatures or other influences, transform into small adults; sometimes these adults are less than half the normal size. Larvae that feed on dark brood combs transform into moths that may be dark gray to almost black; those that feed on foundation become silvery-white moths that are smaller than those that fed on brood comb.

The female starts depositing eggs from four to ten days after she emerges from the cocoon; she continues depositing as long as her vitality lasts. Egg laying may be rapid at times; females have been known to deposit more than 100 eggs in one minute. The total number of eggs laid by a female varies considerably, but it is usually fewer than 300. The adults may live as long as three weeks.

Other Comb-Damaging Moths

The lesser wax moth *(Achroia grisella)* also does some damage to stored combs. Its larvae inflict damage similar to that of the greater wax moth larvae, but the tunnels are smaller, the webs finer, and feeding and webbing are more confined to the outer surface of the combs.

The larvae of the Mediterranean flour moth *(Ephestia kuehniella)* feed on pollen in the hive, and do some damage to combs by boring tunnels through the midrib. The flour moth larvae also tunnel into brood cells and consume the food intended for the developing bee larvae.

These two moths may be controlled by the methods used for control of the greater wax moth.

Control Measures in the Apiary

The most effective natural enemies of greater wax moths are the bees themselves. When the colony is strong, the bees will carry the moths out of the hive and prevent any damage by the larvae. It is only when the colony has been weakened—by disease, starvation, or other means—that the wax moth succeeds in seriously damaging inhabited combs. Accidental loss of queens late in the fall may mean the loss of colonies from greater wax moth damage before the first spring inspection.

Therefore, any beekeeping practices or manipulations that help maintain strong colonies will also decrease the chances of greater wax moth infestation. There is no better insurance against the ravages of the pest than to have strong, queenright colonies.

There is one beekeeping practice that is especially important in preventing greater wax moth infestation—keeping the hive clean. Propolis, burr combs, and refuse should be removed from the bottom board, since they provide protection for larvae of the greater wax moth, even in strong colonies.

Control Measures for Stored Combs

When combs are removed from the hive and placed in storage, there is increased danger of damage by the greater wax moth. Steps must be taken to kill any existing stages of the greater wax moth and guard against later infestation. The most satisfactory method of controlling the insect in stored combs is the use of fumigants and the proper storing of the combs after fumigation.

FUMIGATION OF STORED COMBS

Fumigants are liquid or solid chemicals that form gases when exposed to the air. These gases kill the adult moths, larvae, pupae, and sometimes the eggs. If the combs are thoroughly aired after fumigation, fumigants do not leave residues that would be harmful to bees.

Paradichlorobenzene and ethylene dibromide used as directed effectively protect stored combs from the greater wax moth. They are not recommended for use on combs containing honey intended for human consumption.

Fumigation with paradichlorobenzene. Paradichlorobenzene (PDB) is a white crystalline substance that evaporates slowly in air. It is most effective at temperatures above 70° F. and volatilizes more rapidly as the temperature rises. The gas is heavier than air, nonflammable, and nonexplosive.

PDB can be used to protect all combs in storage except those containing honey intended for human consumption. The odor of PDB is readily absorbed by honey, and though the bees do not object to this odor, such honey is unfit for market purposes. Stored honey combs protected with PDB can be used for spring feeding as long as the combs are aired for a few hours before being placed on colonies.

Treatment with PDB requires no special storage facilities. The supers should be stacked as tightly as possible, and special precautions should be taken to see that the gas, which is heavier than air, cannot escape at the bottom of the stack. For long periods of storage, as over winter, the cracks between supers should be covered with strips of gummed paper. No more than five full-depth supers or ten half-depth supers should be used in a stack. Taller stacks may not allow for complete diffusion of the heavy gas, especially during periods of low temperature.

In a stack of five ten-frame supers, three ounces of the crystals (six tablespoons) should be sprinkled on the frames of the top super. The crystals may be placed directly on the top bars of the frames, or, preferably, separated from the frames by a piece of paper or cardboard. The cover should then be put tightly in place.

At intervals of two or three weeks the covers of the stacks should be raised and the top supers examined; if crystals are no longer present, more should be added. PDB kills adults and immature stages, but not eggs. The continuous presence of crystals within the stack not only repels moths, but also kills any young larvae that hatch after the combs are placed in storage.

Fumigation with ethylene dibromide. Ethylene dibromide is sold as a heavy, clear liquid. It is nonexplosive, nonflammable, and easily stored. On exposure to air it forms a colorless gas that is heavier than air and has a slight, not unpleasant odor. This compound volatizes and diffuses rather slowly, killing all stages of the greater wax moth, including the egg.

Except when an especially prepared fumigation chamber is available, fumigation should take place out-of-doors, under an open shed, or in a well-ventilated room away from workrooms or workers. As in PDB fumigation, supers should be stacked so that the bottom of the stack and the cracks between supers are as gas-tight as possible. No more than eight full-depth supers should be placed in a stack.

An absorbent material such as paper towel, cloth, or sacking is placed on the top bars of the frames in the top super, and sprinkled with one tablespoon of the liquid fumigant. The cover is then quickly put in place. If the temperature is below 60° F., or if the supers are not tightly stacked, two tablespoons per stack of eight supers may be used.

For commercial fumigation of many supers in a relatively airtight room, ethylene dibromide should be used at the rate of two pounds per 1,000 cubic feet of storage space. The liquid should be sprinkled on an absorbent material placed on top of the stacks, as described above; but in an airtight chamber, covers are not necessary, and stacks may be open at the bottom and slightly raised off the floor to promote circulation of the fumigant.

Fumigation with ethylene dibromide should continue for at least twenty-four hours, since the gas diffuses slowly, especially at temperatures below 60° F.

Fumigation with carbon dioxide. Carbon dioxide can be used as a fumigant to destroy all stages of the moth on comb honey or stored combs. Since a high concentration of carbon dioxide is required for fumigation, a relatively airtight room is necessary.

A fumigation period of four hours is required with a concentration of carbon dioxide of 98 percent by volume, at a temperature of 100° F., and a relative humidity of 50 percent. *Do not enter the fumigation chamber while fumigating.* Bee equipment fumigated with carbon dioxide does not require aeration to dissipate residues.

STORAGE

The threat of damage by the greater wax moth to stored combs is continuous, except when temperatures in the storage area drop below 40° F. The presence of PDB in the stacked supers throughout the storage period is a constant check on the greater wax moth; but ethylene dibromide provides better temporary control. Fumigants may effectively destroy all existing stages of the moth, but they may not prevent a later reinfestation.

Therefore, fumigation with these chemicals should be followed by storing the frames in a moth-free room that is clean, well lighted, and ventilated. The supers should be placed on end and spaced to allow air circulation. This will help repel greater wax moths, which like to lay their eggs in dark, poorly ventilated places. The common method of storing combs in tightly closed, crowded hive bodies is highly favorable for wax moth infestation and development.

NONCHEMICAL CONTROL

Temperature extremes can be used to control this pest, because the growth and development of the greater wax moth depends on temperature. Use of high or low temperatures avoids the hazard of honey contamination.

Heat. All stages of the greater wax moth are killed at a temperature of 115° F. for eighty minutes. At a higher temperature, 120° F., the

SUMMARY OF CONTROL METHODS
OF THE GREATER WAX MOTH

Treatment	Empty comb	Full comb [1] (processed)	Full comb [2] (nonprocessed)
Paradichlorobenzene	Yes	No	No
Ethylene dibromide	Yes	Yes (tolerance 125p.p.m.)	No
Carbon dioxide	Yes	Yes	Yes
Heat	Yes	No[5]	No[5]
Cold	Yes	Yes	Yes

---

[1] Honey to be used for human consumption after extraction and processing.
[2] Honey to be used for human consumption without extraction and processing.
[3] Honey to be used for bee food. (Not for human consumption.)
[4] Paradichlorobenzene is most effective at temperatures above 70° F. Eggs of the greater wax moth are not destroyed by PDB.
[5] Not recommended because the temperatures required may cause sagging of combs.

time of exposure can be reduced to forty minutes. Make sure you allow combs to reach the required temperature before measuring the exposure time. *Warning:* Be careful not to expose honeycombs to temperatures in excess of 120° F.

If heat is used to control the greater wax moth, follow these precautions:

• Heat treat only those combs having little or no honey (wax softened at high temperatures would seriously weaken the combs).

• Heat treat supers of the combs when in an upright position — not on the ends or sides.

• Provide adequate air circulation so that the heat will be uniformly distributed throughout the comb. (Ventilating fans are useful for this purpose.)

• Turn the heater off and allow combs to cool before moving the supers.

Cold. Low temperatures can also be used to destroy all stages of the greater wax moth. Use of low temperatures avoids the sagging problem which sometimes occurs when combs are heat-treated. Thus, combs

Full comb[3] (bee food)	*Temperature or dosage level*	*Length of exposure*
Yes	3 ounces or 6 tablespoons per stack of 5 supers.	Keep constant supply while in storage.[4]
Yes	1 tablespoon per stack of 8 supers or 2 pounds per 1,000 cubic feet.	24 hours at a minimum of 60° F.
Yes	98 percent by volume at 100° F.	4 hours.
No[5]	115° F.	2 hours.
Yes	20° F. 10° F. 5° F.	4.5 hours. 3 hours. 2 hours.

with honey and pollen can be cold-treated without much danger to the combs. The minimum temperature and exposure time to destroy all stages of the greater wax moth follow:

Temperature (° F.)	*Time in hours*
20	4.5
10	2.0
5	2.0

The use of heat or low temperatures avoids the hazard of residues, and bee equipment can be reused without endangering the honeybees.

In most cases, honey houses can be converted into a heating chamber by adding a thermostat and a circulating fan. Home freezers can be used for the cold treatment—the only limitation being the size of the freezer.

Once the combs are treated, they should be stored in a chamber which prohibits the entry of adult greater wax moths. Combs should be inspected monthly for signs of infestation, especially if temperatures rise above 60° F.

Precautions

The fumigants ethylene dibromide and PDB used improperly can be injurious to man and animals. Follow the directions and heed all precautions on the container label. Keep the fumigants well labeled, in a dry place where they will not contaminate food or feed and where children and pets cannot reach them.

Ethylene dibromide requires special care in handling. If the liquid is spilled on the skin, it causes blisters or burns if not washed off immediately. The gas is irritating to the lungs and nasal passages. PDB fumes in high concentrations may be irritating to the eyes and nasal passages.

Fumigate out-of-doors, or under an open shed, or in a well-ventilated room away from workrooms or workers.

Where large quantities of combs or equipment must be fumigated, wear a gas mask; have someone work with you or watch you; expose yourself as little as possible; and post warning signs to prevent accidental exposure of others.

Do not use ethylene dibromide or PDB on stored combs containing honey intended for human consumption.

CHAPTER 18

Pesticides

by Philip F. Torchio, *apiculturist, Entomology Research Division, Agricultural Research Service*

One of the major problems faced by beekeepers in the United States and in most other countries with highly developed agriculture is the poisoning of bees by pesticides.

By nature, bees from a colony roam and visit flowers over an area of several square miles. The intensity of visitation in any one part of the area is determined by the relative attractiveness of the flowers. The extent of damage to the colony caused by a pesticide application is affected by the number of bees from a colony working in the treated area, the type of food (nectar and pollen) collected, the time of day the pesticide is applied, the method and number of applications, and the relative toxicity of the material to bees.

The damage can occur when the bee collects food from treated or drift-contaminated plants or by contact with the pesticide on the plant or in the air. The brood can be damaged if fed contaminated pollen.

Reprinted in cooperation with Utah Agricultural Experiment Station from *Beekeeping in the United States*, USDA Agriculture Handbook 335.

Symptoms of Pesticide Poisoning

The following are some usual symptoms of pesticide poisoning. Not all of them are likely to be seen at any one time. Nor are they conclusive as pesticide poisoning, as they can also be the result of other causes. None of these symptoms indicate which material may have affected the bees.

1. An excessive number of dead bees in front of the colony.

2. An unusual number of dead colonies at one time, particularly if they contain honey.

3. A depleted population when the colony should be strong.

4. Sudden cessation of food storage.

5. Dead or deserted brood, with honey in the hive.

6. Dead bees on the floor of the hive during mild weather.

7. A severe break in the brood rearing cycle.

8. A cessation in flower visitation (of significance especially where pollination is desired).

9. Bees crawling from the entrance to die nearby.

10. Dead bees in the hive—on the tops of frames or on the bottom board.

11. The absence of the usual "hum" of workers in the air.

12. Incoming nectar- or pollen-laden bees attacked at the hive entrance by other bees.

13. An unusual number of bees emerging from the entrance carrying dead bees. (The normal daily death rate inside the colony is about 100 bees.)

14. Paralyzed, stupefied, or preening bees on weeds or other objects in the apiary.

Chemical Analysis of Bee Samples

Establishing proof of bee poisoning by chemical analysis is difficult even when the type of pesticide is known. Many pesticides break down rapidly when exposed to the elements. For a sample to be of any value to the analytical chemist, it should be collected immediately after exposure and kept under deep-freeze conditions until analyzed.

There is no federal laboratory equipped for routine analysis of bee samples for all pesticide residues. Some state experiment stations are equipped to determine certain residues. If analysis of a sample of bees is desired, the state experiment station or extension service should be consulted before the sample is submitted to determine whether an analysis can be made. Some commercial laboratories analyze for residues on a fee basis.

Dead bees at hive entrance, a symptom of pesticide poisoning. (USDA photograph)

Determining Toxicity Levels of Pesticides

The problem of determining toxicity levels on honeybees in the laboratory and field has been studied for years by many federal and state research workers. Hundreds of pesticide formulations have been tested in both dusts and sprays. These tests *have not been* conducted on low-volume or ultra-low-volume applications. Limited experience indicates that with these methods of application the hazard to bees increases with certain materials. Extra precautions should be taken until the hazard of pesticides applied by these new methods has been determined. The results of field and laboratory tests *using diluted* pesticides are incorporated in Table 1.

TABLE 1.
SUMMARY OF TOXICITY AND POISONING
HAZARD OF PESTICIDES TO HONEYBEES[1]

Pesticide[2]	Type[2]	Laboratory toxicity	FIELD Toxicity
Abate	P		
aldrin	C	Very high	Very high
allethrin	B	Low	
Aramite	M	do	Moderate
azinphosethyl	P	Very high	
azinphosmethyl	P	do	
Banol	Ca		
Bay 39007	Ca	High	
Bay 39007 G	Ca		Low
Bay 41831	P	Very high	
Bidrin	P	do	
binapacryl	D	Low	
Bomyl	P	Very high	
calcium arsenate	I	High	Very high
carbaryl	Ca	Low–high	High
carbaryl G	Ca		Low
carbophenothion	P	Moderate	High
chlorbenside	M	Low	
chlordane	C	Very high	High–very high.
chlorobenzilate	M	Moderate	
chloropropylate	M		
Chlorthion	P		
Ciodrin	P	Very high	
cryolite	I	High	High
DDT	C	Moderate	Moderate–high
demeton	P	Very high	
diazinon	P	do	Very high
dicapthon	P	do	
dichlorvos	P	do	
dicofol	M	Low	
dieldrin E	C	Very high	
dieldrin G	C		Moderate
dieldrin WP	C	Very high	Very high

[1] Arrangement and much of data from Johansen, C.A., "Summary of the Toxicity and Poisoning Hazard of Insecticides to Honeybees," *Gleanings Bee Cult.* 94: 474-475, 1966.

[2] Chemical names given at end of section on Pesticides.

[2] A=acetamide, B=botanical or derivative, C=chlorinated hydrocarbon, Ca=carbamate, Co=carbonate, D=dinitro compound, I=inorganic compound, M=specific miticide, and P=organophosphorus compound.

[3] Classification of materials: I=hazardous to bees at any time, II=not hazardous if applied when bees are not foraging, III=not hazardous to bees at any time.

APPLICATION IN DUST		FIELD APPLICATION IN SPRAY		
Residual effect	*Use class*[3]	*Toxicity*	*Residual effect*	*Use class*[3]
		Low	3 hours	II
	I	Very high		I
				III
	II-III	Low		III
		Very high	1 day +	I
		do	2–4 days	I
		do	> 7 hours	I
		High	> 1 day	I
	III			
				I
		Very high	5 hours–1 day +.	I
		Low	< 2½ hours	III
		Low–high	2 days	I
Long	I			I
3 days +	I	Moderate–high.	7–12 days +	I
< 2 hours	III			
> 1 day	I	High	< 5 hours	II
		Low	< 2 hours	III
2–3 days	I	High		I
		Low		III
		do	< 1 day	III
				I
				I
	I	High		I
2–3 days	I-II	Moderate	1 day +	II
		do	< 3 hours	II
1 day +	I	Very high	1 day	I
				I
		Very high	1 day +	I
		Low		III
		High	2 days	I
< 2 hours	II			
8 days	I	Very high	5–7 days	I

TABLE 1.
SUMMARY OF TOXICITY AND POISONING
HAZARD OF PESTICIDES TO HONEYBEES[1]   (Continued)

Pesticide[2]	Type[3]	Laboratory toxicity	FIELD Toxicity
Dilan	C	Low	Low–high
dimethoate	P	Very high	
dimetilan	Ca	Moderate	
Dimite	M	Low	
dinitrobutylphenol	D	Very high	
dinitrocresol	D	High	
dinocap	D	Low	
dioxathion	P	do	
disulfoton	P	Very high	
Di-Syston G	P		Low
DN–111	D	Low	
endosulfan	C	Moderate	
endrin	C	Very high	
EPN	P	do	High
ethion	P	Low	
famphur	P	Very high	
fenson	M		
fenthion	P	Very high	
Genite 923	M	Low	Moderate
heptachlor	C	Very high	
heptachlor G	C		Moderate
Hooker HRS–16	M		
Imidan	P	Very high	
isobenzan	C	do	
isodrin	C	Moderate	
Isolan	Ca	High	
isopropyl parathion	P	Low	
Kepone	C	do	
Kepone bait	C		Low
lead arsenate	I	Very high	
lime sulfur	I	Low	
lindane and benzene hexachloride.	C	Very high	Very high
malathion	P	do	do
malathion G	P		Low
Matacil	Ca	High	
menazon	P		
methoxychlor	C	Low	
methyl demeton	P	High	
methyl parathion	P	Very high	
Methyl Trithion	P	do	
mevinphos (Phosdrin)	P	do	Very high

180

Residual effect	Use class[4]	Toxicity	Residual effect	Use class[4]
3 hours	II	Low–high	3 hours	II
		Very high	1–2 days	I
		Low	3 hours	II
		do		III
		Very high	1 day +	I
		do		I
				III
		Low–high	2 hours	II
		Low	3 hours	II
< 2 hours	III			
				III
		Low	< 5 hours	II
		Moderate	< 2 hours	II
1 day +	I	Very high		I
		Low–high	< 2 hours	II
				I
		Low	< 2 hours	III
		Very high	2–3 days +	I
	II–III	Low	< 2 hours	III
		Very high		I
< 2 hours	II			
		Low	< 1 day	III
		Very high	1–4 days	I
		High	> 2 hours	II
				II
		Low	3 hours	II
				III(?)
		Low	< 1 day	III
	III			
		Very high	Long	I
		Moderate		III
2 days +	I	High		I
1 day +	I	Moderate–very high.	2 hours–2 days +.	I
None	II			
		Very high	> 3 days	I
		Moderate	< 2 hours	II
		do	< 1 day	II
		do	None	II
				I
		High	< 1 day	I
	I	Very high	2 hours–1 day.	I

181

TABLE [1].
SUMMARY OF TOXICITY AND POISONING
HAZARD OF PESTICIDES TO HONEYBEES[1]   (Continued)

Pesticide[2]	Type[3]	Laboratory toxicity	FIELD Toxicity
mirex G	C		Low
Morestan	Co		do
naled E	P	Very high	
naled WP	P		High
Naugatuck D–014	M		
Nemacide	P	High	
Neotran	M	Low	
nicotine sulfate	B	do	Low ?
Nissol	A		
ovex	M	Low	Low–high
oxydemetonmethyl	P	High	
paraoxon	P	Very high	
parathion	P	do	Very high
Perthane	C	Moderate	Moderate
phorate E	P	do	
phorate G	P		Moderate
phosphamidon	P	Very high	
Phostex	P	Moderate	
propyl thiopyrophos-phate (NPD).	P	Very high	
Pyramat	Ca	do	
pyrethrum	B	Low	Low
ronnel	P		
rotenone	B	Low	Low–high
ryania	B	do	
schradan	P	Low–very high.	
sodium hexafluorosili-cate bait.	I		Low
Strobane	C	Low	
sulfotepp	P	Very high	
sulfur	I	Low	
Sulphenone	M	Moderate	Moderate
TDE	C	do	do
Temik G	Ca		Low
tepp	P	Very high	Very high
tetradifon	M	Low	
Tetram	P	do	
Thiocron	P	High	
thioquinox	Co	Moderate	
toxaphene	C	Low	Low–high
trichlorfon	P	Low–high	High
Zectran	Ca	Very high	
Zinophos	P	do	

182

Residual effect	Use class[4]	Toxicity	Residual effect	Use class[4]
	III			III
None	III	Low	None	III
		Very high	3 hours	II
> 7 hours	I	do	> 3 hours	II
		Low	< 3 hours	III
		do	2 hours	II
		do		III
Few hours	III			
		Low	3 hours	II
	II–III	do		
		Moderate	None	II
				I
1 day +	I	High	1 day +	I
1 day +	II	Low	< 1 day	II
		Very high	5 hours	II
< 2 hours	II			
		Very high	2 hours–2 days	I
		High	2 hours	II
		Low	2½ hours	II
				I
3 hours	III	Low		III
		Moderate	3 hours	II
< 1 day	II–III			
		Moderate	> 3 hours	II
		Low	1 day	III
	III			
				III
				I
	III	Low		III
	II–III	do		III
	II	Moderate		II
None	III			
< 3 hours	II	Very high	3 hours	II
		Low	< 2 hours	III
		Moderate		II
		Low	3 hours	II
				II
< 1 day	I–II	Low	< 1 day	II
> 3 hours	I	Low–high	2–5 hours	II
		Very high	1–2 days	I
				I

Wild bees are also damaged not only by contaminated food but also by contaminated leaf material, florets, soil, and other articles used in nesting. Furthermore, pesticide residues in soil can remain toxic to ground-nesting bees for several years. Table 2 shows the comparative toxicity of twenty-nine pesticides to the alkali bee and the leafcutting bee.

Laboratory tests of pesticide formulations on honeybees determine toxicity levels on individuals but do not indicate seriousness of damage to the field force and its pollinating or honey-production potential. The field testing of insecticides on the honeybee is especially difficult, because individuals visit fields briefly, the social organization normally prevents all members of the colony from being equally exposed to pesticides, and a part of the field force of any colony may be visiting areas outside the confines of the experiment. Any or all of these factors can drastically affect results. Furthermore, there is no accurate measurement of sublethal effects on a colony exposed to a pesticide application.

Techniques are used that determine (1) toxic effects of direct applications, by placing caged bees in the field as the field is being treated; (2) fuming effect, by placing additional caged bees in the field at hourly intervals after treatment until mortality ceases; and (3) residual effects on the field force, by counting floral visitors before and after exposure to treated and untreated areas for a particular period. In addition, the examination of colonies in treated and untreated areas and counting dead bees in front of hives before and after applications provide a basis for evaluating the effect of the material on the colony. The combined data obtained from these testing techniques provide for an intelligent estimate of pesticide effects on honeybees in the field.

Reduction in Bee Losses

Observation of the following precautions can significantly reduce bee losses from pesticide poisoning.

GROWER COOPERATION

The grower should use a pesticide only when needed. The benefit of the material should outweigh the harm it does the bees. The value of the bees as pollinators should be considered, as well as the effect of the pesticide on them. The effect of the pesticide on the pollinators of other crops in the area should also be considered. A pesticide aiding one crop could seriously reduce production of another one in the area.

TABLE 2.
COMPARATIVE TOXICITY OF 29 PESTICIDES
TO ALKALI BEE AND ALFALFA
LEAFCUTTING BEE[1]

Pesticide	Alkali bee	Alfalfa leafcutting bee
Aramite	III	
Bidrin	I	I
carbaryl	I	I
carbophenothion	II	I
DDT	III	I
demeton	III	III
diazinon	I	
dicofol	III	III
dimethoate	I	I
dioxathion		III
endosulfan	I	I
endrin	II	I
EPN	I	
ethion		I
malathion	II	I
menazon		III
methoxychlor	III	
mevinphos	I	
naled	II	II
oxydemetonmethyl	III	
parathion	I	I
phosphamidon	I	I
Phostex		III
ronnel		I
schradan		III
tepp	III	
tetradifon	III	
toxaphene	III	I
trichlorfon		I

[1] Classification of materials: I=hazardous to bees at any time, II=not hazardous if applied when bees are not foraging, III=not hazardous to bees at any time.

Select the right pesticide. All pesticides are not equally hazardous to bees. Some pesticides will kill an entire colony, some will weaken it, but still others are safe. Select the pesticide that is least hazardous to pollinators and that will control the harmful pests.

Apply granules or sprays rather than dust. Granules are, in general, harmless to bees. Sprays drift less than dusts.

Use ground equipment. Airplanes discharge pesticides at higher altitude and with greater turbulence than ground machines. This increases the likelihood that bees in flight will come in contact with the pesticide or that it will drift onto adjacent crops or into apiaries. Time the pesticide application. The safest time to apply pesticides is when bees are not working plants. Treat at night or at a time of day when bees are not in the field.

Avoid drift of the pesticide. Bees cluster on the hive entrance on hot days and nights where they can be exposed to drifting pesticides. Wait until the night is sufficiently cool for the bees to move inside. Colonies can be damaged by fumes of some pesticides, such as parathion, azinphosmethyl, malathion, and benzene hexachloride. Notify the beekeepers near areas to be treated so that they may move or otherwise protect the colonies. However, notification is not a release of responsibility.

Confine bees to hives when hazardous pesticides are to be applied. At left, photos show hive being covered with burlap. Above, beekeeper soaks burlap with water. Bees should be covered at night, when the bees are in the hives, and the burlap should be soaked at least once every hour. Confinement should be as short as possible.

BEEKEEPER COOPERATION

Select safe bee locations. Place colonies away from agricultural areas if possible, away from fields routinely treated, or at least where they will not be subject to drift of the material from the treated field.

Identify the colonies. Post owner's name, address, and telephone number in a conspicuous place in the apiary. Let the nearby growers know where the bees are located so the beekeeper can be notified.

Know the pesticides. Be acquainted with pesticides likely to be used in the area and their potential hazard to bees.

Confine the bees when hazardous materials are applied. Beehives can be covered with plastic sheeting that will confine the bees and exclude pesticide sprays, dusts, or fumes. Since heat builds up rapidly under plastic exposed to the sun, confinement can only last for a few hours after dawn on warm days. This may be long enough to protect the bees from some materials.

Hives can also be covered with wet burlap for a day or more, even during the hottest weather, and the bees will not suffer from lack of air or water. The hives should be covered at night when all the bees are in the hives. During the day the burlap should be soaked with water at least once every hour.

Relocate the colonies if they are likely to be repeatedly exposed to hazardous pesticides.

Bees are valuable to the grower. Try to convince him of their value to him and of the importance of protecting them.

CHAPTER 19

Commercial Beekeeping Equipment

by Charles D. Owens and Benjamin F. Detroy, *agricultural engineers, Agricultural Engineering Research Division, Agricultural Research Service*

To be successful in his business of producing bulk honey, a beekeeper must be an efficient manager. He must make sure that the honey house is designed for the work to be done in it, and that it is properly equipped. By wise selection of commercially available equipment, he can reduce labor, production time, and costs.

Because of differences in the size of beekeeping enterprises and in details of operation, beekeepers have individual problems. Some may need equipment that is not available from suppliers. Usually it is possible for them to build the equipment themselves.

Equipment For The Honey House

The main items of equipment needed in the honey house are those used in handling supers, uncapping, extracting, cleaning and clarifying, heating and cooling, wax and honey separation, and bulk handling of honey.

From *Selecting and Operating Beekeeping Equipment*, Farmers' Bulletin 2204.

A two-wheeled handtruck is a versatile piece of equipment around the honey house. (USDA photograph)

HANDLING SUPERS

Moving supers. Numerous devices are available for moving supers in the honey house. They include:

- Handtrucks. (Some handtrucks also function as lifting devices.)
- Motorized trucks.
- Dollies with casters.

A handtruck can be used to move single stacks of supers. The supers should be stacked on skid boards so they can be picked up with the truck.

If you have large numbers of supers to be moved, a motorized truck may be warranted. Motorized trucks are available in several sizes.

Dollies with casters supplement a truck. Supers are unloaded from the truck to the dollies and moved to the uncapping machine.

Dollies with casters are useful for moving filled supers in the honey house. (USDA photograph)

Dollies or skid boards used in a hot room should have an open plat-form to allow circulation of air through the stacks. An open platform can be made so that the super rim fits on the frame. If supports are provided under the frame, a metal sheet can be slid under the frame to form the drip pan. When the stacks are in the hot room, the drip pan should be removed to allow circulation of air through the stacks.

Pallets and drip trays. Equipment for handling pallets can be obtained from suppliers of beekeeping equipment. This equipment is for single stacks. Equipment for handling up to four stacks can be obtained from suppliers of industrial trucks.

Skid boards and pallets can be used on the truck and therefore have an advantage over dollies, which are difficult to hold on the truck without special devices.

Drip trays are useful in the honey house and on trucks. In the honey house they are placed under supers to keep honey from dripping on the floor. On the bed of a truck, they keep the bed clean. If you move stacks one at a time with a handtruck, make a tray to hold individual stacks. If you move them with a warehouse truck, make trays that will each hold two or four stacks.

Lift tables. A considerable amount of labor can be saved at the un-capping center if lift tables are used to keep the supers at a convenient working level at all times. Supers are usually stacked on these tables so that the top one is at working height. After the top super has been emptied and removed, the stack is raised until the next super is at the desired height. The process of removing the top super and raising the stack is continued until the last super in the stack is emptied.

Lifting units of various types are available from suppliers of beekeeping equipment. At least one model can be used as a handtruck.

UNCAPPING

Heated knives, vibrating knives, and semiautomatic machines have been developed to help beekeepers reduce the labor involved in uncapping combs. The more elaborate uncapping machines cost several hundred dollars each.

The extent to which a beekeeper should invest in uncapping equipment depends on the scope of his operation, but the need for efficiency is especially urgent at this early stage of production. Even after prudent investment in an uncapping device—for example, an electrically heated hand knife—a beekeeper is likely to find that uncapping is the most laborious, most time-consuming step in the extracting operation.

Most beekeepers use a hand knife or a hand plane for uncapping. Knives and planes operate more efficiently when heated. Models equipped for automatic heating are on the market. Some are heated by electricity, some by steam. The operator holds a comb in one hand and with the other hand draws the heated knife or plane over the face of the comb.

A power-operated vibrating knife is fastened to a frame by spring steel mounts. It may be mounted in whatever position the operator desires—vertically, horizontally, or in an inclined position. The knife, which in all commercial models is steam heated, vibrates in the direction of its length. To remove cappings, the operator draws the face of the combs across the knife.

Several types of semiautomatic uncapping machines are available. Some are equipped with rotary knives and some with vibrating knives that cut in a straight line. Rotary knives use force, not heat, to cut the cappings and to keep the knives free of cappings. Vibrating knives are heated.

Frames are uncapped mechanically after being fed into the machine by hand. They are then delivered to a collecting unit or are removed from the machine by hand. If frames are delivered to a collecting unit, the uncapped combs are carried on a set of chains and allowed to drain. When the number of combs on the chains is sufficient to load an extractor, they are transferred to the extractor by hand.

A power uncapper makes the uncapping task less fatiguing; it may

This horizontal vibrating uncapping knife is power-operated. Knives are usually steam-heated. (USDA photograph)

not increase the speed with which it is accomplished. An unskilled operator can uncap faster with a machine than by hand. But a person who is skilled in uncapping by hand can accomplish as much—until he becomes fatigued—as he can with a machine. A person uncapping by hand must have considerable manual strength to keep pace with a power uncapper over a long period. Thus, machines are preferred if the uncapping is done by women.

EXTRACTING

After combs are uncapped, they are placed on a holding device or delivered to a collecting unit.

The device for holding the combs could be a reel or a rack over a tank, which may be part of the uncapping machine or part of the unit that holds the cappings. In either case, the honey drippings from the uncapped combs are collected and delivered to the sump. (In a small operation, it is satisfactory to place the combs directly into an extractor after uncapping. This procedure is uneconomical in larger operations, because it requires two extractors—while one is extracting, the other is being filled. The second extractor is an unnecessary expense.)

There are two types of extracting machines—the radial and the reversible basket. Both extract the honey by centrifugal force.

A radial extractor starts at a speed of 150 r.p.m. and reaches a maximum speed of 300 r.p.m. Both sides of the comb are extracted simultaneously and the extracting cycle requires twelve to twenty minutes.

This is a semi-automatic un-capper. After frames of honey are inserted between the guide fingers, they are automatically pulled downward between two vibrating uncapping knives. Uncapped frames are then conveyed over a drip tank. (USDA photograph)

Radial extractors range in size from those that accommodate twelve frames per load to those that accommodate fifty frames.

A reversible-basket extractor is equipped with at least two baskets that support the combs during the extracting cycle. During the extracting cycle, centrifugal force acts on one side of the comb, then on the other; the baskets are reversed three or four times—turned 180°—until both sides of the comb are completely extracted. The extracting time ranges from two to four minutes at a constant speed.

Reversible-basket extractors range in size from those that accommodate two frames per load to those that accommodate sixteen frames. In most situations, the four- or eight-frame size is preferred. An extractor of more than eight-frame size is undesirable in a portable extracting plant because of the limited space.

Theoretically, it is necessary to have only three supers in a portable plant at one time if a reversible-basket machine is used, whereas six to fifteen supers are necessary for continuous operation of a radial machine. Fewer supers in the plant means that space is made available for other extracting equipment or for honey storage.

Automatic controls relieve the operator of all tasks except loading, starting the machine, and unloading. These controls change the extractor speed during the extracting cycle, reverse the basket (reversible-basket extractor), and shut off the motor when the cycle is completed.

A small commercial extracting setup consists of a four-frame reversible-basket extractor with automatic control, melter with heat lamps, and sump with float control for the pump motor. (USDA photograph)

SUMP AND PUMP

Honey flows from the uncapping and extracting operations into a collecting tank, called a sump. The main function of the sump in a continuous-flow system is to maintain a constant supply of honey for the conditioning equipment. The extractor cannot perform this function, because the flow of honey from the extractor is irregular.

The sump also acts as a collecting unit in other systems. The pump is started when the sump is full and continues to operate until the honey is down to the pump intake. The pump motor is shut off and is idle until the sump is again full.

A sump usually contains a series of screens or baffles (or both) for removing most of the coarse particles of wax and other foreign material from the honey. It may have an exterior water jacket in which the temperature of the water is maintained at 120° to 140° F. to facilitate pumping or gravity flow.

In a very simple production setup, a pump is not necessary—gravity performs its functions. But in most honey houses, a pump is necessary for moving honey through the heating and cooling units, and may be needed in straining; the role of gravity is limited to filling and emptying storage tanks.

The gear pump and the vane pump are the most common types. If the pump is part of a continuous-flow system, it should be supplied with honey in sufficient quantities to allow uninterrupted operation. Other systems require pump operation only long enough to move a given quantity of collected honey.

To prevent introduction of air into the honey, run the pump at low speed and keep the level of the honey in the sump well above the pump intake.

The sump should be equipped with a float switch that will start and stop the pump motor automatically.

A continuous-flow system should include a pressure switch in the honey line. Its purpose is to stop the pump motor if excessive pressure develops in the line. High line pressure should be avoided in the interest of protecting equipment in the conditioning system.

CLEANING

To meet U.S. Department of Agriculture grade A standards, extracted honey must be free of foreign particles that are removable by a standard No. 80 sieve at a honey temperature of 130° F.

Honey may be cleaned by flotation or by straining. In the flotation process, liquid honey is pumped, or flows by gravity, into settling tanks. Particles of wax and other foreign material less dense than the honey rise to the top of the tank and are skimmed off; the honey is then drawn from the bottom of the tank. When drawing off honey, the operator should leave the top layer in the tank; it may contain foreign particles.

Strainers are available from suppliers in a wide variety of shapes and sizes, and many types have been constructed by beekeepers to meet their individual needs. The honey is moved through the strainer by pressure (pumping) or by gravity flow.

Ease of cleaning the straining material is an important consideration in selecting a strainer. The material may be metal screen, layers of silica sand, or cloth.

To be efficient, screens must be of fine mesh. They are very difficult to clean when they become plugged with particles of wax.

Silica sand is also difficult to clean. Another disadvantage is that after the sand is washed it contains a large amount of water. Before the sand is reused, the water must be removed to keep it from being added to the honey.

Cloth is easier to clean than the other materials, and since the initial cost is low it may be discarded when cleaning becomes difficult. The chief disadvantage of cloth is its inability to withstand strong pressure. Cloth requires more frequent cleaning when used in a pressure unit than when used in a gravity unit; the cleanings are necessary to prevent an increase in line pressure that would rupture the cloth.

In one type of strainer the straining material consists of a cylindrical rotating metal screen, one end of which is slightly elevated. Honey and wax are fed into the cylinder at the elevated end. Wax particles adhere to each other and form a roll, which helps to knead the honey through the screen. The honey is collected in a tank beneath the screen, and the wax is discharged from the lower end of the screen cylinder. The manufacturer guarantees an output of 500 pounds of honey per hour if a thirty-mesh screen is used and if the honey being strained was uncapped with a hand knife or vibrating knife. With a fifty-mesh screen, output is reduced by one-half.

Where consistently large quantities of honey are to be strained, continuous operation is desirable. This is made possible by alternate operation of two strainers; while one is idle for cleaning, the other is in service.

HEATING AND COOLING

Straining is facilitated if the honey temperature is between 100° and 120° F. But honey ahead of a strainer should not have a temperature above 120°. Such a temperature softens the wax particles, and soft particles are hard to remove with a strainer because they are forced into or through the small openings. After the honey has been strained, it must be heated to a higher temperature to prevent fermentation and to retard crystallization.

Improper heating with insufficient cooling causes honey to darken and impairs the flavor. It is especially important to guard against overheating.

Before honey is strained, it can be heated by passing it through a double-jacket tank (with baffle), or by passing it over a heated shallow pan with corrugated bottom, or by passing it through a heat exchanger designed for heating honey. The first method of heating is used most extensively. If the honey is to be strained in a continuous-flow system, the heat exchanger should be used.

Any of the three types of heaters used before straining may be used afterwards to apply additional heat to prevent fermentation and to retard crystallization.

Concentric tube heat exchangers and gravel strainers used in a continuous-flow honey-conditioning system. (USDA photograph)

A batch heater of the jacketed tank type may also be used for heating honey after straining.

Honey must be stirred while heating or the honey next to the walls will overheat and burn while that in the center will still be cold. Stirring can be performed by hand, but a slow-speed motor-driven agitator is best. (It is always necessary to stir honey that is held in a container for heating.)

Heat exchangers of modern design have tubes containing a honey channel three-sixteenths inch thick. The channel is between two layers of flowing hot water. Honey is pumped through the channel in a direction opposite to the direction of the flow of water. These units can be connected in series so that the honey is in the heating channel for sufficient time to raise it to the desired temperature. The temperature of the water supply should be maintained at 180° to 190° F. to raise the honey temperature quickly with a minimum number of heat exchangers.

Four procedures for heating liquid honey, and for cooling it, are suggested on page 198. Follow the procedure that is best suited to your equipment.

1. Heat rapidly to 160° F. Total heating time should not exceed 10 minutes. Heat exchangers should be used. Begin cooling immediately after heating; cool to 100° or less; complete cooling in five minutes or less.

2. Heat to 140° in not more than ten minutes; hold thirty minutes; cool to 100° or less in not more than ten minutes. (If heating requires more than ten minutes, decrease hold time accordingly.)

3. If cooling facilities are not available and heated honey is to be placed in containers of less than 60-pound capacity, heat to 120° to 130°. The time required for heating is not critical.

4. If cooling facilities are not available and heated honey is to be placed in 60-pound cans, in drums, or in large tanks, heat to 120° or less. The time required for heating is not critical.

In both heatings (before and after straining), the heating agent should be hot water—not steam. With hot water as the agent, it is easy to control the amount of heat added; with steam, overheating would be a danger.

The hot water may be introduced into the heating units from electric water heaters or from a central heating plant, or electric immersion heaters may be installed in the units.

Beekeepers must devise their own cooling systems, or strive to get maximum performance from those that they purchase; no fully satisfactory equipment is available. In small operations, cooling pans and devices similar to heating units, but containing cold water, are practicable aids in cooling honey.

RECOVERING HONEY FROM CAPPINGS

Cappings accumulated in the uncapping of combs contain honey, and the beekeeper has the task of recovering the honey without impairing its flavor or color. The following methods are used:

1. Draining, followed by melting or pressing. The combs are uncapped into wire baskets, or into perforated sheet-metal baskets, which hold the cappings until most of the honey is drained off.

After the cappings have drained, some honey remains in them; this is recovered by melting or pressing the cappings. The drained cappings may be melted in a solar melter or double boiler, or pressed with a wax press. If reinforced perforated sheet-metal baskets are used, pressure can be applied to the cappings with a hydraulic or mechanical jack.

2. Flotation. A type of cappings melter available from suppliers of beekeeping equipment separates honey from cappings by flotation. The cappings containing honey enter the melter tank beneath a steam-heated coil. The honey and wax separate by gravity. The separation is facilitated by heat, which softens the cappings and increases the fluidity of the honey. The wax, being less dense than the honey, rises to the top where it is melted by heat from the steam coil. The honey level, which is controlled by an adjustable height overflow enclosed by a baffle to prevent the entry of wax, is maintained at least two inches below the steam coil. A layer of wax in various stages of liquefaction is maintained between the honey level and the top of the steam coil by a wax discharge opening. As the wax from the cappings rises toward the steam coil from underneath, liquid wax flows out the discharge into a solidifying container. The honey flows out the overflow into a pipe leading to the honey sump.

Slumgum (residue consisting of cocoons, propolis, and other foreign material) should be removed regularly; it acts as an insulator and reduces the effectiveness of the heating coils.

3. Centrifuging. Honey is separated from cappings by centrifugal, or whirldry, extractors. These are radial extractors that have reels with solid bottoms and perforated steel sides. Centrifugal separating units can be purchased, or radial extractors can be converted to centrifugal units by fitting them with wire baskets into which the cappings are placed.

Centrifuging is usually faster, and produces drier cappings, than draining, but after each one it is necessary to melt the wax to salvage the remaining honey. Honey obtained by melting the wax after draining or centrifuging (or by flotation if proper precautions are not taken) is usually dark and off flavor. It may be used for spring feeding of bees where spreading of bee diseases is not a problem.

A person who wishes to use the flotation method and does not have a regular source of steam heat for heating the coils can melt the cappings with electric heat lamps. These lamps should be placed seven inches above the wax level in the melter tank. When the melter is started in the morning, it is necessary to melt a solidified wax bed. For this, five or more watts per square inch of melter surface are required. For melting wax after the solidified wax bed has been melted and while uncapping is in progress, two or three watts per square inch of melter surface are required.

This solar wax melter is easy to construct and extremely efficient. The amateur beekeeper will find it a handy device.

Solar wax melter. A solar wax melter, for melting cappings and old combs, is useful in any honey house, and it eliminates the fire hazard that is present when wax is melted by direct heat.

Melters of this type range in size from 1½ by 3 feet to 2½ by 14 feet.

The following suggestions are for beekeepers who wish to make solar wax melters:

- The melter should be large enough to accommodate the volume of wax that would be placed in it in one day.

- It should be made of aluminum, stainless steel, or copper sheet, to prevent darkening the wax.

- It should be shallow—about five inches deep—to prevent shadows on the melting area.

- It should have a double-glass cover, and the space between the two pieces of glass should be one-half inch to one inch.

- It should be tight enough to keep out bees.

- The sides and bottom should be insulated.

- The drain should be screened to prevent slumgum from leaving the melter with the honey and molten wax.

Wax press. Wax presses are used to recover wax from slumgum. Since presses are among the more expensive pieces of beekeeping equipment, a small operator may decide to have his wax rendering done by a commercial plant. But if he makes this decision, he should melt down old combs and cappings as they accumulate. A solar wax melter is recommended for this purpose. If the combs and cappings are not melted down, they may be destroyed by wax moths.

Do not discard the slumgum. At least half of its weight is wax, and commercial rendering plants can recover the wax at a low cost.

STORAGE AND SHIPMENT

After honey has been conditioned, it is run into large storage tanks, which may also be used as settling tanks. From the storage tanks it is transferred to containers for shipment to the packer.

Containers may be five-gallon (sixty-pound) cans or fifty-five-gallon drums. Choice between cans and drums depends on the facilities that you have for moving containers and on the floor space available for them before they are shipped.

For moving a large number of cans, you need caster dollies, hand trucks, or a motor-driven lift truck. If drums are used, it will probably be necessary to stack them to conserve floorspace; for this you will need a hoist or a motor-driven lift truck. A barrel truck (a hand truck of special design for handling barrels) is satisfactory for moving drums across the floor, but not for stacking them.

Motor-driven lift truck is moving sixteen sixty-pound cans of honey. Attachments for handling barrels and drums are available. (USDA photograph)

Glossary

Abdomen. Segmented posterior part of bee containing heart, honey stomach, intestines, reproductive organs, and sting.

Acarapis woodi (Rennie). Scientific name of acarine mite, which infests tracheae of bees.

Acarine disease. Disease caused by acarine mite.

Alighting board. Extended entrance of beehive on which incoming bees land.

American foulbrood (AFB). Contagious disease of bee larvae caused by *Bacillus larvae* White.

Antennae. Slender jointed feelers, which bear certain sense organs, on head of insects.

Anther. Part of plant that develops and contains pollen.

Apiarist. Beekeeper.

Apiary. Group of bee colonies.

Apiculture. Science of beekeeping.

Apis. Genus to which honeybees belong.

Apis dorsata Fabricius. Scientific name for giant bee of India; largest of all honeybees.

Artificial cell cup. *(See* Cell cup.)

Artificial insemination. Instrumental impregnation of confined queen bee with sperm.

Bacillus larvae White. Bacterial organism causing American foulbrood.

Balling a queen. Clustering around unacceptable queen by worker bees to form a tight ball; usually queen dies or is killed in this way.

Bee bread. Stored pollen in comb.

Bee dance. Movement of bee on comb as means of communication; usually same movement is repeated over and over.

Bee escape. Device to let bees pass in only one direction: usually inserted between combs of honey and brood nest when removal of bees from honey is desired.

Bee gum. Usually hollow log hive; occasionally refers to any beehive.

Beehive. Domicile prepared for colony of honeybees.

Bee louse. Relatively harmless insect that gets on honeybees, but larvae can damage honeycomb; scientific name is *Braula coeca* Nitzsch.

Bee metamorphosis. Stages in development of honeybee from egg to adult.

Bee moth. *(See* Wax moth.*)*

Bee paralysis. Condition of bee, sometimes caused by virus, that prevents it from flying or performing other functions normally.

Bee plants. Vegetation visited by bees for nectar or pollen.

Bee space. Amount of space acceptable to bees, neither too narrow nor too wide; discovered by great American beekeeper Langstroth.

Beeswax. Wax secreted from glands on underside of bee abdomen; molded to form honeycomb and can be melted into solid block.

Bee tree. Hollow tree in which bees live.

Bee veil. Screen or net worn over head and face for protection from bee stings.

Bee venom. Poison injected by bee sting.

Bee yard. *(See* Apiary.*)*

Bottom board. Floor of beehive.

Brace comb. Section of comb built between and attached to other combs.

Braula coeca Nitzsch. *(See* Bee louse.*)*

Breathing pores. *(See* Spiracles.*)*

Brood. Immature or developing stages of bees; includes eggs, larvae (unsealed brood), and pupae (sealed brood).

Brood chamber. Section of hive in which brood is reared and food may be stored.

Brood comb. Wax comb from brood chamber of hive containing brood.

Brood nest. Area of hive where bees are densely clustered and brood is reared.

Brood rearing. Raising bees.

Bumble bee. Large hairy bee in genus *Bombus.*

Burr comb. Comb built out from wood frame or comb, but usually unattached on one end.

Cap. Covering of cell.

Capped brood. *(See* Sealed brood.)

Capped honey. Honey stored in sealed cells.

Carniolan bee. Gentle grayish-black bee originally from Carniolan Mountains in or near Austria.

Caucasian bee. Gentle black bee originally from Caucasus area of Russia; noted for its heavy propolizing characteristic.

Cell. Single unit of space in comb in which honey is stored or bee can be raised; worker cells are about 25 cells per square inch of comb, drone cells are about 18 per square inch.

Cell cup. Queen cell base and part of sides; artificial cell cups are about as wide as deep.

Chilled brood. Immature stages in life of bee that have been exposed to cold too long.

Chunk comb honey. Type of honey pack in which piece of honeycomb is placed in container of liquid honey or wrapped "dry" in plastic container.

Circadian rhythm. Biological rhythm with period length of about 1 day.

Clarified honey. Honey that has been heated, then filtered to remove all wax or other particles.

Cleansing flight. Flight bees take after days of confinement, during which they void their feces.

Clipped queen. Queen whose wing (or wings) has been clipped for identification purposes.

Cluster. Collection of bees in colony gathered into limited area.

Colony. Social community of several thousand worker bees, usually containing queen with or without drones.

Comb. *(See* Honeycomb.)

Comb foundation. Thin sheet of beeswax impressed by mill to form bases of cells; some foundation is also made of plastic and metal.

Comb honey. Edible comb containing honey; usually all cells are filled with honey and sealed by bees with beeswax.

Commercial beekeeper. One who operates sufficiently large number of colonies so that his entire time is devoted to beekeeping.

Cross-pollination. Transfer to pollen from anther of one plant to stigma of different plant or clone of same species.

Crystallization. *(See* Granulated honey.)

Cut-comb honey. Comb honey cut into appropriate sizes and packed in plastic.

Demaree. Method of swarm control, by which queen is separated from most of brood; devised by man of that name.

Dequeen. Remove queen from colony.

Dextrin. Soluble carbohydrate of poor nutritive value to bee.

Dextrose. Also known as glucose; one of principal sugars of honey.

Diastase. Enzyme that aids in converting starch to sugar.

Division board. Flat board used to separate two colonies or colony into two parts.

Division board feeder. Feeder to hold sirup; usually size of frame in hive.

Drawn comb. Foundation covered with completed cells.

Drifting bees. Tendency of bees to shift from their own colony to adjacent ones.

Drone. Male bee.

Drone brood. Area of brood in hive consisting of drone larvae or pupae.

Drone comb. Comb having cells measuring about four to the inch and in which drones are reared.

Drone egg. Unimpregnated egg.

Drone layer. Queen that lays only infertile eggs.

Dwindling. Rapid or unusual depletion of hive population.

Dysentery. Unusual watery discharge of bee feces, often associated with nosema disease.

Emerging brood. Young bees first coming out of their cells.

Enzyme. Material produced by both man and animals that acts on another material to change it without changing itself.

Escape board. Board with one or more bee escapes on it to permit bees to pass one way.

European foulbrood. Infectious disease of larval brood, caused by *Streptococcus pluton* (White).

Excluder. *(See* Queen excluder.)

Extracted honey. Honey extracted from comb.

Extractor. Machine that rotates honeycombs at sufficient speed to remove honey from them.

Feces. Bee droppings or excreta.

Fecundate. To inseminate or implant sperm into female.

Fertilize. To make fertile, as by implanting sperm into ova.

Field bees. Bees 2½ to 3 weeks old that collect food for hive.

Flash heater. Device for heating and cooling honey within few minutes.

Food chamber. Hive body containing honey-filled combs on which bees are expected to live.

Foulbrood. Common name of two brood diseases; usually applied to American foulbrood.

Foundation. (See Comb foundation.)

Frame. Wood case for holding honeycomb.

Fructose. (See Levulose.)

Fumagillin. Antibiotic given bees to control nosema disease.

Galleria mellonella (L.). Scientific name of greater wax moth.

Giant bee. (See Apis dorsata Fabricius.)

Glucose. (See Dextrose.)

Grafting. Transfer of larvae from worker cells into queen cells.

Granulated honey. Crystallized or candied honey.

Gynandromorph. Bee having both male and female characters.

Half-depth super. Super only half as deep as standard 10-frame Langstroth super.

Heterosis. Greater vigor displayed by crossbred animals.

Hive. Man-constructed home for bees.

Hive tool. Metal tool for prying supers or frames apart.

Hobbyist beekeeper. One who keeps bees for pleasure or occasional income.

Hoffman frame. Self-spacing wood frame of type customarily used in Langstroth hives.

Honey. Sweet viscous fluid elaborated by bees from nectar obtained from plant nectaries, chiefly floral.

Honeybee. Genus Apis, family Apidae, order Hymenoptera.

Honeycomb. Comb built by honeybees with hexagonal back-to-back cells on median midrib.

Honeydew. Sweet secretion from aphids and scale insects.

Honey extractor. *(See* Extractor.)

Honeyflow. Period when bees are collecting nectar from plants in plentiful amounts.

Honey house. Building in which honey is extracted and handled.

Honey pump. Pump for transferring liquid honey from one container to another.

Honey stomach. Area inside bee abdomen between esophagus and true stomach.

Honey sump. Temporary honey-holding area with baffles; tends to hold back sizable pieces of wax and comb.

Hormone. Substance produced in small quantity in one part of body (usually in gland of internal secretion) and transported to other parts, where it exerts its action.

Hymenoptera. Order to which all bees belong, as well as ants, wasps, and certain parasites.

Introducing cage. Small wooden and wire cage used to ship queens and also to release them quietly into cluster.

Invertase. Enzyme produced by bee that speeds inversion of sucrose to glucose and fructose.

Italian bees. Bees originally from Italy; most popular race in United States.

Jumbo hive. Hive 2½ inches deeper than standard Langstroth hive.

Langstroth frame. 9⅛- by 17⅝-inch frame.

Langstroth hive. Hive with movable frames; each frame usually 9⅛ by 17⅝ inches.

Larva. Stage in life of bee between egg and pupa; "grub" stage.

Laying worker. Worker bee that lays eggs after colony has been queenless for many days.

Legume. One of Leguminosae, or plants such as clover, alfalfa, peas, or beans.

Levulose. Fructose or fruit sugar; one of sugars, with glucose, into which sucrose is changed.

Mandibles. Jaws of insects.

Mating flight. Flight taken by virgin queen when she mates with drone in air.

Metamorphosis. Changes of insect from egg to adult.

Migratory beekeeping. Movement of apiaries from one area to another to take advantage of honeyflows from different crops.

Mite. *(See Acarapis woodi* (Rennie).)

Movable frame. Frame bees are not inclined to attach to hive because it allows proper bee space around it.

Nectar. Sweet exudate from nectaries of plants.

Nectaries. Special cells on plants from which nectar exudes.

Nosema disease. Disease of bees caused by protozoan spore-forming parasite, *Nosema apis* Zander.

Nucleus (nuclei). Miniature hives.

Nurse bees. Young worker bees that feed larvae.

Observation hive. Hive with glass sides so bees can be observed.

Ocellus (ocelli). Simple eye(s) of bees.

Package bees. Screen wire and wood container with 2 or 3 pounds of live bees.

Parafoulbrood. Relatively rare bee disease similar to European foulbrood; caused by bacterium *Bacillus para-alvei* Burnside.

Paralysis. *(See* Bee paralysis.)

Parthenogenesis. Production of offspring from virgin female.

Pheromone. Chemicals secreted by animals to convey information to or affect behavior of other animals of same species.

Pistil. Part of flower extending from ovary to stigma.

Play flight. Short orientation flight taken by young bees, usually by large numbers at one time and during warm part of day.

Pollen. Dustlike material produced in flower and necessary on stigma of female flower for seed production; also collected in pellets on hindlegs of bees.

Pollen basket. Area on hindleg of bee adapted for carrying pellet of pollen.

Pollen cake. Cake of sugar, water, and pollen or pollen substitute for bee feed.

Pollen substitute. Mixture of water, sugar, and other material, such as soy flour, brewer's yeast, and egg yolk, used for bee feed.

Pollen supplement. Mixture, usually of six parts (by weight) pollen, 18 parts soy flour, 16 parts water, and 32 parts sugar.

Pollen trap. Device installed over colony entrance that scrapes pollen from legs of entering bees.

Pollination. Transfer of pollen from male to female element of flower.

Pollinator. Agent that transfers pollen.

Pollinizer. Plant that furnishes pollen for another.

Proboscis. Tongue of bee.

Propolis. Resinous material of plants collected and utilized by bees within hive to close small openings or cover objectionable objects within hive.

Pupa. Stage in life of developing bee after larva and before maturity.

Queen. Sexually developed female bee.

Queen cell. Cell in which queen develops.

Queen excluder. Device that lets workers pass through but restricts queen.

Queenless. Without queen.

Queen rearing. Producing queens.

Queenright. With queen.

Queen substance. Material produced from glands in head of queen; has strong effect on colony behavior.

Ripe honey. Honey from which bees have evaporated sufficient moisture so that it contains no more than 18.6 percent water.

Robbing. Bees of one hive taking honey from another.

Royal jelly. Food secreted by worker bees and placed in queen cells for larval food.

Sacbrood. Minor disease of bees caused by filterable virus.

Sealed brood. Brood in pupal stage with cells sealed.

Self-pollination. Transfer of pollen from male to female element within same flower.

Septicemia. Usually minor disease of adult bees caused by *Pseudomonas apiseptica* (Burnside).

Shallow super. Super less than $9\frac{9}{16}$ inches deep.

Shipping cage. Screen and wood container used to ship bees.

Skep. Beehive made of straw.

Smoker. Device used to blow smoke on bees to reduce stinging.

Solar wax extractor. Glass-covered box in which wax combs are melted by sun's rays and wax is recovered in cake form.

Spermatheca. Small saclike area in queen in which sperms are stored.

Spermatozoon. Male reproductive cell.

Spiracles. External openings of tracheae.

Stamen. Male part of flower on which pollen-producing anthers are borne.

Stigma. Receptive part of style where pollen germinates.

Sting. Modified ovipositor of female Hymenoptera developed into organ of defense.

Streptococcus pluton (White). Causative agent of European foulbrood.

Sucrose. Cane sugar; main solid ingredient of nectar before inversion into other sugars.

Super. Extra division of hive above brood nest area.

Supersedure. Replacement of one queen by another while first is still alive.

Swarm. Natural division of colony of bees.

Tarsus. Fifth segment of bee leg.

Thorax. Middle part of bee.

Tracheae. Breathing tubes of insects.

Tumuli. Nest mounds.

Uncapping knife. Knife used to remove honey cell caps so honey can be extracted.

Unite. Combine one colony with another.

Unsealed brood. Brood in egg and larval stages only.

Virgin queen. Unmated queen.

Wax glands. Glands on underside of bee abdomen from which wax is secreted after bee has been gorged with food.

Wax moth. Lepidopterous insect whose larvae destroy wax combs.

Wild bees. Any insects that provision their nests with pollen, but do not store surplus edible honey.

Winter cluster. Closely packed colony of bees in winter.

Wired foundation. Foundation with strengthening wires embedded in it.

Wired frames. Frames with wires holding sheets of foundation in place.

Worker bee. Sexually undeveloped female bee.

Worker comb. Honeycomb with about 25 cells per square inch.

Worker egg. Fertilized bee egg.

Additional Sources
of Information

BEE JOURNALS

There are several magazines devoted to the subject of beekeeping. Some of these are listed below. The publishers are generally glad to send you a free sample copy upon request.

American Bee Journal, Hamilton, Illinois 62341

Canadian Beekeeping, Box 128, Orono, Ontario, Canada L0B 1M0

Gleanings In Bee Culture, The A.I. Root Co., P.O. Box 706, Medina, Ohio 44256

The Speedy Bee, Rt. 1, Jesup, Georgia 31545

BEEKEEPERS' ORGANIZATIONS

Most states have a state beekeepers' association. Some also have county or regional organizations. Ask your County Agricultural Extension Agent for information regarding these. Their meetings give you a chance to hear speakers on various aspects of beekeeping. Also, you have a chance to associate with other beekeepers and to ask questions and compare notes. You will find it very worthwhile to join your local organization.

BULLETINS ON BEEKEEPING

The Cooperative Extension Service of many State Land Grant Colleges publishes bulletins on beekeeping. These are helpful for acquainting you with local conditions that affect beekeeping practices. Write to your State Extension Service for information on these or contact your County Extension Agent.

CORRESPONDENCE COURSES

Correspondence courses in beekeeping are available in British Columbia, Minnesota, Nebraska, New Jersey, New York, Nova Scotia, Ohio, Pennsylvania and Tennessee.

SHORT COURSES

Annual University short courses in beekeeping are held in British Columbia, Florida, Maryland and Pennsylvania.

BOOKS

Dadant & Sons, ed. *Hive and the Honeybee.* rev. ed., Dadant and Sons, Hamilton, Ill. 62341, 1975.

Dadant, C. P. *First Lessons in Beekeeping.* Journal Printing Co., rev. ed., Carthage, Ill. 62341, 1968.

Eckert, John E., and Frank R. Shaw. *Beekeeping.* MacMillan Co., New York, 1960.

Kelley, Walter T. *How to Keep Bees and Sell Honey.* Walter T. Kelley Co., Clarkson, Ky., 1973.

Morse, Roger A. *Complete Guide to Beekeeping.* E. P. Dutton & Co., Inc., New York, 1972.

Root, A. I. *ABC and XYZ of Bee Culture.* 33rd. ed., A. I. Root Co., Medina, Ohio, 44256, 1966.

Root, John A. *Starting Right With Bees.* 14th ed., A. I. Root Co., Medina, Ohio, 44256, 1967.

Taylor, Richard. *How-to-do-it Book of Beekeeping.* Walnut Press, Naples, N.Y., 14512, 1974.

Other Garden Way Books
You Will Enjoy

Perhaps you are interested in something bigger than bees for your homestead. If so, you will find some Garden Way books that will offer you information to help you to decide which animals you might like to raise, and to give good guidance on the hows and whys of keeping these animals.

Here are some of those books:

Raising Rabbits the Modern Way, by Robert Bennett. 156 pp., quality paperback, $3.95. Everything for the home and semi-commercial producer.

Raising Poultry the Modern Way, by Leonard Mercia. 240 pp., quality paperback, $4.95. Outstanding in this field.

The Family Cow, by Dirk van Loon. 270 pp., quality paperback, $5.95. Invaluable for the owner of a single cow or small herd. We recommend this highly.

Raising Sheep the Modern Way, by Paula Simmons. 240 pp., quality paperback, $5.95. The definitive book on small-scale sheep-raising.

Raising a Calf for Beef, by Phyllis Hobson. 128 pp., quality paperback, $4.95. Not only how to buy and raise your calf, but complete instructions for butchering and cutting beef for the home freezer.

Raising Milk Goats the Modern Way, by Jerry Belanger, editor of *Countryside & Small Stock Journal.* 152 pp., quality paperback, $3.95. Belanger knew the problems goat owners faced, the questions they wanted answered. This book gives the answers in down-to-earth, easy-to-understand terms.

These books are available at your bookstore, or may be ordered directly from Garden Way Publishing, Dept. BK, Charlotte, Vermont 05445. If order is less than $10, please add 60c postage and handling.

Index

A

Acarine disease, 160
Acetic acid, 158
Adult bee, 92; anatomy, 111-13;
  development of, 95-96; digestion,
  113-14; diseases, 156-60; foods, 115-
  18; nutritional requirements, 114-
  15; prevention of disease, 160-61
American foulbrood, 10, 145-50; char-
  acteristics, 155; checking for, 72-73
Amoeba disease, 159-60
Anatomy, 111-13
Apiary, 88
Artificial diets for bees, 117-18
Ascosphaera apis, 154
Aspergillus, 154

B

Bacillus larvae, 145, 150
Balling a queen, 110
Bee behavior, 97-110
Beebread. 84
Bee brush, 6
Bee cluster, 91-92, 124
Bee colony, dividing, 53, 98-99;
  morale, 103; moving, 74; odor, 104;
  queenless, 138-39; queenright, 138-
  39; seasonal activity, 90-96; social
  pattern, 98
Bee escape, 6, 60-61
Bee-louse, 160
Bee needs, 84
Bees, how many to order, 23; how to
  handle, 12-22; how to obtain, 2;
  installation of in hive, 26-33; when to
  order, 1-2
Beeswax, 10, 69, 89, 95-96, 112-13;
  how made, 69
Bee yard, 88

Boardman entrance feeder, 41
Bottom supering, 56, 127
Brace comb, 100
Braula coeca, 160
Brood, 34, 39, 85, 91; cells for, 100;
  dead, 145; diseases, 145-54; food, 92;
  rearing, 37, 91-92, 101-2, 121-22,
  123, 125, 126; sealed, 17, 39, 52, 145;
  with fungus diseases, 154
Brood chamber, 72, 85; reversing in
  spring, 42, 126-27, 128
Brood comb, 99-100; infected with
  American foulbrood, 146-47;
  worker, 99-100
Brood diseases, 145-54
Brood nest, 33, 126, 137-38; tempera-
  ture, 3
Bulk comb honey, supering for, 54-58
Bulk honey production, equipment,
  188-201

C

Capping tub, 8, 66, 68
Carbon dioxide, 171, 173
Carmelization, 70
Carrots for Children (recipe), 135
Cell cappings, 8, 65, 95; methods for
  recovering honey from, 198-201;
  removing, 67
Chalkbrood, 154
Checking hives, 21-22, 32-33
Cleansing flights, 74
Clothing, 5-6, 12-14
Cluster, 72, 91-93, 98, 124
Cold knife, 65-66
Collecting unit, 192
Comb, 21, 56, 59-61, 69; brace, 100;
  building, 99-100; storage by bees,
  100; storing of, 171; wax, 95-100
Crops for bees, 83, 84

D

Dadant hive, 88
Dances, 105
Deep super, 4, 7; adding, 33
Digestion, 113-14
Diseases, adult bee, 156-60; brood, 145-54; fall check for, 72-73; fungus, 154; multiple, 154; non-infectious, 155-56 (table); prevention of, 160-61
Dividing a colony, 53; by bee consensus, 98-99
Division-board feeder, 42
Dollies, 189-90
Doolittle, G.M., 97
Double graft, 138
Drip tray, 190
Drone (s), 39, 47, 52, 90-94, 139; activities of, 109; comb, 52, 99; development of, 96

E

Egg laying, 39, 50, 91-93
Eggs, 17
Entrance, 31, 33, 51; block, 31, 51, 75; feeder, 30-32, 41; for winter, 73, 75, 123
Equipment, 2-11; for extracted honey production, 8-9, 188-201; how to sterilize, 11; sources of, 9-10; used, 10-11
Escape for bees, 6, 60-61
Ethylene dibromide, 170-71, 173, 174
European foulbrood, 145, 150-52; characteristics of, 155
Extracting machines, different types, 192-93
Extractor, 8-9, 65-68, 191-93, 199

F

Feeder block, 31
Feeders, different types, 30-32, 41-42
Feeding, 20; fall, 72-73; late winter, 125; spring, 37-39, 42; winter, 122-23
Flight behavior, 100
Flour moth, 168
Food chamber, 4, 72; reversing in spring, 42

Food requirements, 115-18
Food supply, storage, 5, 37, 42, 72, 90, 92, 94-95, 122
Forager bee, 105
Foraging, 105-9
Foundation, 7, 33, 58-59, 89; different types for comb honey production, 56
Frames, 55-56, 58-60, 123; uncapping, 191
Fumagillin, 73; to prevent nosema, 157-58
Fumidil-B, 35, 158
Fumigants, 169-71; precautions, 173-74
Fungus diseases, 154

G

Greater wax moth, 162-68; control of, 168-74
Guard bees, 43

H

Hand plane for uncapping, 191
Handtrucks, 189
Heaters for honey, different types, 196-97
Heat exchangers, 196-98
Hive(s), checking, 21-22, 31-33, 34; Dadant, 88; different types, 85-89; entrance for winter, 73; extra, 7; Langstroth, 85-89; location of, 25-26, 73-74, 80-84; moving, 74; opening of, 16-22; packing for winter, 42, 76-79, 123-25; preparation of, 25; temperature control, 103, 124-25; ventilation, 51-52; weak, 72
Hive bodies, 123; reversing position of, 42
Hive management, 119-30; fall and winter, 71-79, 122-25; late winter, 125; spring, 36-53, 125-26; summer, 54-70
Holding tank, 68-69
Honey, 50, 69, 84; cleaning, 195-96; crystallized, 68-70; equipment for bulk production, 188-201; for winter stores, 72; how made, 69; in cooking, 131-32; recipes, 133-36; storage comb, 100; storage of, 201; straining, 68-69, 195-96; to prevent

Honey (cont.)

crystallization, 196; to prevent
fermentation, 69-70, 196
Honey, bulk comb, 54-56, 64
Honey, cut-comb, packaging, 61-64
Honey, extracted, 54, 58-60, 64-70;
equipment for production, 8-9, 188-
201
Honey-Baked Chicken (recipe), 135
Honeybee, anatomy, 111-13; digestion,
113-14; foods, 115-18; nutritional
requirements, 114-15
Honey Brunch Cocoa (recipe), 136
Honey Cereal (recipe), 133
Honey Corn Muffins (recipe), 133
Honeydew, composition of, 115
Honey extractor, 8-9, 65-68, 191-93,
199
Honey Gingerbread (recipe), 134
Honey house, equipment for, 188-201
Honey Nut Bread (recipe), 135
Honey-Peanut Butter Cookies (recipe),
134
Honey storage tank, 9, 68
Honey temperature, 196-97; different
procedures for heating and cooling,
198
How many to order, 23; installation in
hive, 26-33
How to obtain, 2

K

Knife (ves), cold, 65-66; different
kinds, 191; thermostatically-
controlled electric, 8, 65-66;
vibrating, 191

L

Langstroth hive, 85-89, 97
Larva, 95; with American foulbrood,
146-48; with European foulbrood,
151-52; with fungus diseases, 154;
with sacbrood, 153
Lesser wax moth, 168
Lift tables, 190-91

M

Malpighamoeba mellificae, 156, 160
Management of hives, 119-30; fall and
winter, 71-79, 122-25; late winter,
125; spring, 36-53, 125-26; summer,
54-70
Mediterranean flour moth, 168
Melter, 198-200; solar wax, 200
Mice, prevention of, 75-76
Motorized trucks, 189

N

Nasanov gland, 103-4
Nectar, 50, 69, 84; composition of, 115
Needs of bees, 84
Non-infectious diseases, 155-56 (table)
Nosema, 35, 52, 121, 156-58
Nosema apis, 156
Nurse bees, 94, 95-96, 156; transmitting
American foulbrood, 148; trans-
mitting sacbrood, 153

O

Oxytetracycline, 150, 152

P

Package bees, how many to order, 23;
installing, 26-33; production of, 140-
43; queen cage, 26-29; receiving, 23-
24; shipping cage, 26-29
Pallets, 190
Paradichlorobenzene, 169-70, 171, 173,
174
Paralysis, 158-59
Pepperbox pattern in brood comb, 147,
151
Pesticides, 175-87; hazard to alkali bee
and alfalfa leafcutting bee, 185
(table); hazard to honeybee, 178-183
(table); precautions, 184-87;
symptoms of poisoning, 176
Pheromones, 50, 97-98, 100-101, 103-4
Plant juices, composition of, 116

Pollen, 84, 90, 140-41; composition of, 116-17 (table); supplements, 125-26; trap, 125
Pollen collecting, control of, 108-9
Pollination, 107-9
Power uncapper, 191-92
Propolis, 10, 84, 91, 160
Pseudomonas apiseptica, 159
Pumps, 194; different types, 195
Pupa, 95; infected with American foulbrood, 147-48; infected with fungus diseases, 154
Purple brood, 156

Q

Queen, 20, 33, 50, 71-72, 85, 90-95, 120-21, 130; activities of, 109-10; cage for shipping, 26-29, 140; development of, 96; diseases of, 35; egg laying, 39, 50, 91-93, 95, 110; finding in spring, 43; rearing commercially, 138-41; rearing of, 137-38; replacement, 44-47, 120-21; supersedure, 34-35
Queen, new, 93-94; introducing, 46
Queen, old, disposing of, 45-46; signs of failing, 39
Queen bank, 140
Queen cells, 20-21, 35, 93, 100, 110; destroying, 52-53, 127; for queen production, 137-41; in preparation for swarming, 47-48
Queen excluder, 4, 56, 58, 128, 130; remove for winter, 73
Queenless colony, 138-39
Queenright colonies, 138-39
Queen substance, 50, 96, 98, 104, 112

R

Radial extractor, 192-93
Recipes, 133-36
Replacement queens, 44-47, 120-21
Requeening, 44-47, 71-72, 120-21; to eliminate European foulbrood, 152; to eliminate paralysis, 159
Reversible basket extractor, 192-93

Robber bees, 31, 38, 43-44; to discourage, 44; transmitting American foulbrood, 148; transmitting sacbrood, 153
Royal jelly, 93, 95-96, 110, 112; composition of, 116
Rules on how to work with bees, 12

S

Sacbrood, 152-154; characteristics of, 155
Scent gland, 103-4
Scout bees, 48, 93, 98
Sealed brood, 17, 39, 52, 145
Septicemia, 159
Shaker box, 143
Shallow supers, 4-5, 7, 54-55, 58, 85, 88
Skid board, 189-90
Slumgum, 199-201
Smoker, 14-16, 21; temper control, 102-3
Starvation, 91-92
Sting(s), how to deal with, 22; pheromone, 104
Stock, 130
Stonebrood, 154
Storage tank for honey, 9, 201
Stored combs, fumigation of, 169-71
Strawberry Special (recipe), 134
Streptococcus pluton, 150, 152
Sucrose, 117
Sugar syrup, 5, 20, 24-25, 31-32, 39; for fall feeding, 72; for late winter feeding, 125
Sulfathiazole, 73, 150
Sump, 192, 194-95
Supering, 54-60, 128; bottom, 56, 58, 127; top, 58
Supers, 85; deep, 4, 7, 33; equipment for moving in honey house, 189-91; for extracted honey, 58-59; returning to hive after extracting honey, 68; shallow, 4-5, 7, 54-55, 58, 85, 88; storage of, 171; when to add, 33, 50, 56-58
Supersedure, 34-35; cells, 53
Swarm box, 138
Swarm cells, 35

Swarming, 21, 47-48, 93-95; causes of, 100-101; control of, 52-53, 126-27; hiving, 48; prevention of, 50-53, 94; signs of, 93

130; when, 122; with laying workers, 47

V

Veil, 14
Ventilation, 51, 74, 114
Vibrating knife for uncapping, 191

T

Temperature control, 51-52, 91-92, 94, 103, 124-25
Temperature extremes to control greater wax moth, 171-72, 173
Top supering, 58
Trucks for handling supers, 189
Two-queen system, 128, 130

W

Water, 52, 82, 84, 92, 114
Wax combs, 95, 99-100
Wax moth, 168
Wax press, 201
Weak hive(s), 72; protecting, 43-44; uniting, 38
Winter packing for hives, 76-79, 123-25; removing of, 42
Worker bees, 52, 84, 85, 90-91, 93, 98, 101; balling a queen, 110; development of, 95-96; laying, 46-47; removing for requeening, 44-45
Worker brood comb, 99-100

U

Uncapper, power, 191-92
Uncapping, 64-68, 191-92; knife, 8, 65-66, 191; machines, 191-92; tub, 8, 66, 68
Uniting colonies, 38, 39, 47, 72, 104,